Building Business Intelligence Using SAS®
Content Development Examples

Tricia Aanderud
Angela Hall

The correct bibliographic citation for this manual is as follows: Aanderud, Tricia, and Angela Hall. 2012. *Building Business Intelligence Using SAS®: Content Development Examples.* Cary, NC: SAS Institute Inc.

Building Business Intelligence Using SAS®: Content Development Examples

Contents

About This Book

The SAS Intelligence Platform provides a toolset that allows organizations to not only reach every organization level, but also to address their most critical business issues. There is a wealth of information freely available to users, everything from simple blog tips to in-depth user documentation. All of this information actually makes it more difficult for users to understand how to get the most power from the toolset, or even know where to start!

Our purpose is writing *Building Business Intelligence with SAS®: Content Development Examples* is to have a single quick-start guide that moves you quickly through each tool so you can learn the basics and then easily expand that knowledge with some intermediate and advanced techniques. We want to ensure that users understand how to get the most out of this software for their organization.

Is This Book for You?

Users of the SAS Business Intelligence or SAS Enterprise Business Intelligence toolsets from all experience levels will find this book useful. In each chapter, there is a tool overview, a step-by-step process for creating output, tips and tricks from tool experts, and ways to enhance the output. At the end of each chapter, you will learn some tool administration information to round out your tool knowledge.

If you are studying for the Business Intelligence Content Developer certification, you will find this book to be a valuable study resource.

Prerequisites

You do not need any previous experience with the toolset or SAS to use this book. If you have prior coding experience, you might find the stored process chapter easier to understand.

Scope of This Book

This book covers the SAS solutions that include these clients:

- SAS Enterprise Guide (used with SAS BI toolset)
- SAS Stored Processes and Prompting
- SAS Information Map Studio
- SAS OLAP Cube Studio
- SAS Management Console
- SAS Web Report Studio
- SAS Add-In for Microsoft Office
- SAS BI Dashboard
- SAS Information Delivery Portal

Authors' Websites

We will continue to update these books and provide expanded information about the SAS BI clients within our blogs:

BI Notes for SAS BI Users: http://www.bi-notes.com

Real BI for Real BI Users: http://blogs.sas.com/content/bi/

Additional Resources

SAS offers you a rich variety of resources to help build your SAS skills to apply the full power of SAS software. Whether you are in a professional or academic setting, we have learning products that can help you maximize your investment in SAS.

Bookstore	http://support.sas.com/publishing/
Training	http://support.sas.com/training/
Certification	http://support.sas.com/certify/
Knowledge Base	http://support.sas.com/resources/
Support	http://support.sas.com/techsup/
Learning Center	http://support.sas.com/learn/
Community	http://support.sas.com/community/

Comments or Questions?

If you have comments or questions about this book, you may contact the authors through SAS as follows:

Mail:

SAS Institute Inc.
SAS Press
Attn: Tricia Aanderud/Angela Hall
SAS Campus Drive
Cary, NC 27513

E-mail: saspress@sas.com

Fax: (919) 677-4444

Please include the title of the book in your correspondence.

SAS Publishing News

For a complete list of books available through SAS Press, visit support.sas.com/publishing.

Receive up-to-date information about all new SAS publications via e-mail by subscribing to the SAS Publishing News monthly eNewsletter. Visit support.sas.com/subscribe.

Acknowledgments

The first acknowledgement goes to Lisa Erwin, who so many years ago as our extraordinary SAS account manager introduced us to this terrific software and believed we could accomplish great things with it. Many people supported the effort behind this book, but none more so than our external review team, Harry Droogendyk, Bob Janka, and Steve Overton. Your candid feedback pointed us in the right direction. There have also been other colleagues, friends and family members, who have assisted in numerous ways. Many, many thanks to all of you who have contributed to making this book a reality.

Tricia Aanderud

Once I heard a motivational speaker who said that in our lives each of us meets at least one *bump person*. Someone who forces us to new levels, reveals our hidden talents, and motivates us to do more than we ever thought possible. Thus, they *bump* us forward. My co-author, Angela Hall, is one of those people in my life. I would never have learned SAS, rather long dreamed of writing a book about business intelligence without a continual *bump* from her.

There were many evenings and weekends spent with my face in a computer. My friends and family (Thanks, Mom!!) were a continual support system during this lengthy process.

However, without my incredible husband Ken, taking care of everything else, my task dedication would not have been possible. Thank you for being the perfect husband—*I love you most of all*.

Angela Hall

After *three* attempts to get this book written it goes without saying that this could not have been done without my co-author Tricia Aanderud. Her creativity, organization, and writing skills were essential to our writing accuracy and speed.

Throughout this process my kids Sydney, Grier, Harrison and Grant continued to be supportive and interested in the book. When I told them I had a sample copy they all came running. They were so excited to finally see and hold what I had been working on many nights and weekends. Your encouragement kept me going. Love all of you dearly!

But none of this would have happened unless I had the most accommodating husband in the world. Craig, your love, support, and understanding make me the person I am today.

Chapter 1

Introduction

Building SAS BI Solutions for Your Organization

Chapter 1

Introduction

All organizations need data not only to survive, but also to compete. How organizations transform that data into actions is basically the definition of business intelligence (BI). The basic questions organizations ask read like a newspaper headline—Who did What, When, Where, and Why. Analytics adds another concept: What will happen next? An organization's ability to ask these questions and receive reliable answers immediately is now paramount to its success.

In some organizations there is a small group who understand the data, where it originates, and how it can be exploited for enterprise-wide benefits. These people are essentially the organizations data gatekeepers. However, different departments have additional information, vantage points, and customer understanding that would further exploit and transform the data into actions. Organizations need a way to place the analytical power of the gatekeepers into a larger audience.

SAS Business Intelligence provides an interface for multiple audiences to dissect, discover, and decide on what the data means. These reporting tools make dynamic information available to all users, giving them the ability to manipulate results and further understand the business. The power of SAS Business Intelligence is in reducing the data gatekeeper role in the organization so each person can interact with analytic results.

SAS offers a BI solution that offers mechanisms to reach each level of the organization. Each toolkit within this product provides a different amount of complexity and functionality to aid a broad deployment. Within this book, the authors have set out to clarify how these products can be fully used to ensure a successful implementation. Within the book, each of these tools is described, providing ideas and examples for use.

1.1 Different Tools for Different People

Essentially, there are three broad groups of SAS BI clients. One group includes the tools required to create reports, the second group allows users to view the reports, and the final group offers the ability to address data management and administration needs.

Typically organizations grant each user community access to one of these three groups; however, you are not restricted to using this grouping in your organization. The authors recommend that you further refine which tools are available to which community of users to successfully meet your organizational requirements and your users' skills and capabilities.

1.1.1 Reporting Tools

Many of the tools provide the ability to create a report and even chose an interface that might be the most familiar to the user. Power users might prefer to use SAS Enterprise Guide and even build stored processes that allow others to view custom reports. With SAS Web Report Studio and SAS Add-In for Microsoft Office, the report creation process can be as simple and familiar as needed.

1.1.1.1 SAS Enterprise Guide

This SAS desktop client provides a rich graphical user interface (GUI) to process, analyze, and report on defined data sources. Users require prior knowledge of SAS or specific SAS Enterprise Guide training to begin to use this product. In the SAS Business Intelligence solution, SAS Enterprise Guide is the recommended interface for SAS Stored Processes authors.

Figure 1.1-1 Example project from SAS Enterprise Guide

1.1.1.2 SAS Add-In for Microsoft Office

After a small download to a user's Microsoft Office installation, the SAS Add-In for Microsoft Office provides another menu item to generate and place analysis reporting results directly within Microsoft Word, Microsoft Excel, Microsoft PowerPoint, and Microsoft Outlook. Within Excel, users can also interact with local or remote data structures. Those who enjoy pivot tables in Excel will especially enjoy the fact that this wizard is the same when accessing data sources made available through SAS.

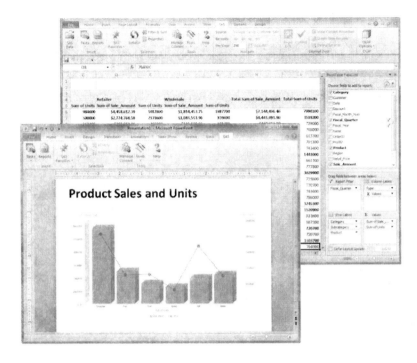

Figure 1.1-2 Example report from SAS Add-In for Microsoft Office

1.1.1.3 SAS Web Report Studio

The easiest reporting tool to learn, this product provides an online ability to generate reports. Features include adding content such as graphs, tables, and cross-tabular summaries; headers and footers; including prompts for user/report interaction; printing to PDF; and scheduling/emailing to groups. This product can exponentially increase the number of report authors within an organization.

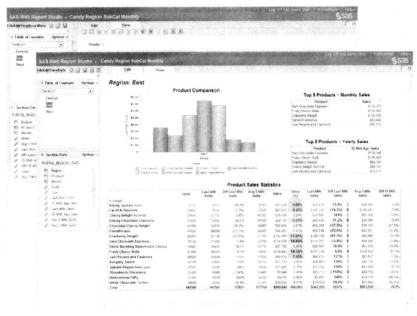

Figure 1.1-3 Creating Web reports through SAS Web Report Studio

1.1.1.4 SAS Stored Processes

You can write SAS code to complete essentially any requirement. In situations when you need a more advanced report than SAS Web Report Studio or SAS Add-In for Microsoft Office can provide, you can write a stored process. Stored processes can be made available for use online, through SAS Web Report Studio, SAS Enterprise Guide, or SAS Add-In for Microsoft Office. For organizations that have existing SAS code, stored processes provide the ability to increase code flexibility and availability.

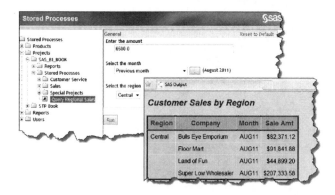

Figure 1.1-4 Custom HTML output through SAS Stored Processes

1.1.2 Viewing Tools

Two tools exist that specifically allow you to display data and reports not only to the user community but also to others in your organization and to those external to your organization.

1.1.2.1 SAS BI Dashboard

Viewing data through the SAS BI Dashboard provides users with a high-level vantage point of the status of their organization. SAS BI Dashboard developers can include the business goals alongside the reports to quickly highlight to users what targets are being met versus those that need immediate attention.

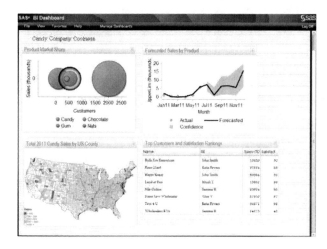

Figure 1.1-5 SAS BI Dashboard sample

1.1.2.2 SAS Information Delivery Portal

Consolidating all of the various reporting outputs available, the SAS Information Delivery Portal offers a central location for viewing information. Reports generated in SAS Web Report Studio, SAS Enterprise Guide, or stored as a stored process are all available for viewing through the SAS Information Delivery Portal.

Figure 1.1-6 Sample SAS Portal page

1.1.3 Administration Tools

Information maps and OLAP cubes are the building blocks of the report-building tools.

1.1.3.1 SAS Information Map Studio

SAS Web Reports must first have data defined to it; this is the job of SAS Information Map Studio. Data administrators can reclassify, sort, and format selected data using this desktop client. New measures can be generated and relational data can be filtered in advance, but filters and prompts can also be generated for report authors to utilize in SAS Web Report Studio at their discretion.

Because information maps are required for SAS Web Report Studio authors, data administrators and report creators must work closely together. A new information map is not required with each new web report.

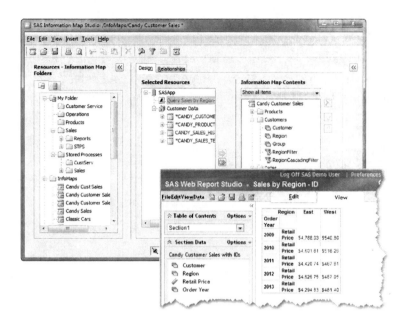

Figure 1.1-7 Sample information map used in SAS Web Report Studio

1.1.3.2 SAS OLAP Cube Studio

Online analytical processing (OLAP) data structures provide an efficient way to store and report measures on large data structures. Essentially cubes of data, the measures are summarized at multiple levels simultaneously. SAS OLAP Cube Studio is a desktop client that offers a GUI to generate, manage, and tune OLAP cubes.

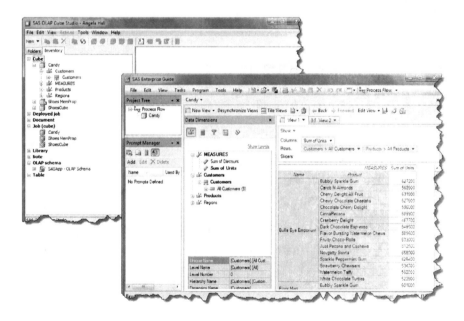

Figure 1.1-8 SAS OLAP Cube Studio and OLAP cube in the OLAP Viewer

1.1.3.3 SAS Management Console

Consolidating content (reports, data tables and libraries, stored processes), server definitions (metadata, OLAP, share, workspace, stored process), and security assignments (users, access) into a single desktop client provides a seamless interface for administrators to manage the system.

Figure 1.1-9 SAS Management Console

1.2 Implementation Decision Streams

Now that you have a better understanding of what these SAS BI client tools provide, you need to map your report requirements with the tools to implement new reports or to move older reports into the new SAS BI environment.

1.2.1 Simple Reports

When reports require no additional manipulation of data or include specific layout and formatting requirements, SAS Web Report Studio is the SAS BI client of choice. Report authors can enable prompts to allow report users to filter the results quickly, without creating duplicate report layouts for the same data tables. Report objects can include data table printouts or graphs (including line plots, and bar and pie charts).

1.2.2 Huge Data Sets

Opening large datasets within any SAS product is dependent on system resources and client PC capabilities. Each query can then reduce the server performance for other users. OLAP cubes summarize data during build time to reduce the response time for queries. SAS Enterprise Guide and SAS Add-In for Microsoft Office users can open OLAP cubes to quickly slice and dice the data to retrieve summary statistics.

1.2.3 Legacy SAS Code and Custom Reports

Quick implementations are realized when legacy SAS code can be converted into SAS Stored Processes. Suppose you have legacy SAS code that is run for a particular product when management requests a report. If you are out of the office, the report cannot be run. You can remove this bottleneck altogether by changing the filter to use a macro variable, converting the code to a stored process, and enabling prompting. Now management can run this stored process immediately from the Web or from SAS Add-In for Microsoft Office when they need it. Management needs only to select the product they are interested in and then run the report.

1.3 Book Organization

The book is organized in such a way to help developers build content and address such implementation decision streams as previously mentioned. The chapters are grouped into the following four sections to quickly provide you with examples and tricks to generate content.

1.3.1 SAS Enterprise Guide, SAS Stored Processes and Prompts

In the following figure, the prompting framework is the underlying key to generating the most power from SAS Enterprise Guide and SAS Stored Processes. Use of prompting is critical to generating significant power from flexible stored process code.

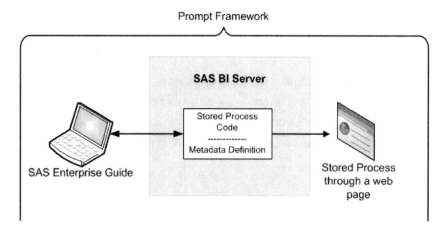

Figure 1.3-1 Prompting framework impact on SAS Enterprise Guide and SAS Stored Processes

1.3.2 SAS OLAP Cube Studio

OLAP cubes can be used as data sources for any other SAS BI clients. Information concerning OLAP cube technology is included in this book to provide users with a basic understanding of what OLAP means and how to use it to address ad hoc query requirements for large data structures.

The OLAP cube represented in the following figure is used for the immediate querying needs of analysts using SAS Enterprise Guide and SAS Add-In for Microsoft Office as well as for multiple SAS Web Report Studio reports.

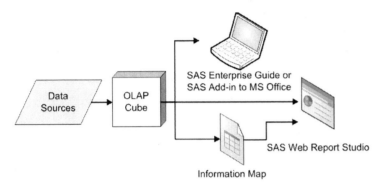

Figure 1.3-2 Multiple purposes for single OLAP cube

1.3.3 Information Maps, SAS Web Report Studio, and SAS Add-In for Microsoft Office

Information maps are the foundation for SAS Web Report Studio reports. Web reports can connect directly to other sources and can include stored process output. However, the majority of implementations use only the information maps within web reports.

SAS Add-In for Microsoft Office provides reporting functionality in excess of the capabilities of SAS Web Report Studio and can retrieve data through information maps or raw tables and OLAP cubes.

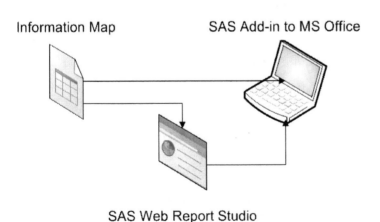

Figure 1.3-3 Information Maps relationship to SAS Web Reports Studio and SAS Add-In for Microsoft Office

1.3.4 The SAS BI Dashboard and SAS Information Delivery Portal

These presentation layers provide a consolidated view for individuals who do not use SAS. The ability to create a central location for these users is paramount to getting business intelligence capabilities across the organization. Information can include links or views of multiple Web reports or graphs based on a variety of data sources. Developers can add functionality for users to interact with dashboards or link into additional research to further analysis the report.

See the example in the following figure. The interactions between all the components can become complex. However if you are a user of BI Dashboard or Portal the end result is simply that all analysis and reporting is available in these locations.

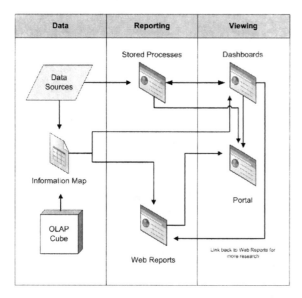

Figure 1.3-4 Sample sources when viewing Dashboard or Portal

1.4 Book Features

As you read this book, there are highlights and topics that make your learning progress quicker.

1.4.1 Quick Tips

Within each chapter, there are quick tips that highlight essential and helpful information to accomplish the task at hand.

 These are denoted by this icon. Use these tips in conjunction with the examples.

1.4.2 Artwork Callouts

Some screenshots have callouts, such as numbers, to assist your understanding of the task steps or to provide a code reference. The callout reference might be in table or might be in line with the text for clarity.

For example, CustID ❶ is related to Customer and ProdID ❷ is related to ProdID.

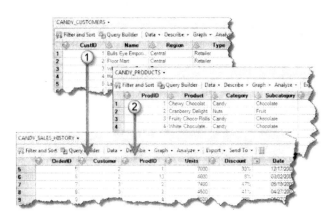

Figure 1.4-1 Example of artwork callouts

1.4.3 Sample Data Sources Used

The examples in this book were created using the sample datasets shipped with the SAS Business Intelligence tool.

- SASHELP Library

  - SASHELP.PRDSAL3

- SAS Enterprise Guide supplied samples located at SAMPLES\DATA where SAS Enterprise Guide is installed.

  - CANDY_CUSTOMERS

  - CANDY_PRODUCTS

  - CANDY_SALES_HISTORY

  - CANDY_SALES_SUMMARY

1.4.4 Sections on Administration Topics

Your organization's SAS administrators use SAS Management Console to set up security, define resources (such as libraries and tables), and manage the systems SAS uses to run the entire solution. This book is aimed at report and content developers; however, to complete some of the examples listed, additional administrator tasks are necessary and are therefore detailed in each chapter in a section titled Administration Topics. Each chapter includes these tips so content developers can provide reference information to SAS administrators.

1.4.5 Applicable SAS Versions

The examples in this book were all built using the following product versions. In the majority of cases, the techniques will be similar to other SAS software versions.

- The third maintenance release for SAS 9.2 (SAS 9.2 TS2M3)
- SAS Enterprise Guide 4.2
- SAS Add-In for Microsoft Office 4.2
- SAS OLAP Cube Studio 4.2

- SAS Information Map Studio 4.2
- SAS Web Report Studio 4.2
- SAS Information Delivery Portal 4.2
- SAS BI Dashboard 4.2

The main difference between SAS 9.2 TS2M3 and prior versions is the new SAS BI Dashboard 4.2, which contains completely new functionality. Therefore, essentially all content in the book, except for the SAS BI Dashboard chapter, can assist users learning how to build SAS Business Intelligence content.

For SAS 9.3 with corresponding 4.3 or 4.3.1 SAS BI client interface environments, all of the content within the book is still applicable.

1.5 Architecture Overview

For business users of SAS Business Intelligence, only a basic understanding of system architecture is required. The following sections describe the key terms that this book uses and that you should understand.

1.5.1 Metadata

Metadata is essentially data about data. Metadata can include pointers to physical locations where the data actually resides, access controls or security information, and an organization structure to quickly find information and how everything runs. Metadata is used by each of the SAS BI clients to grant you the ability to log in and access certain data.

1.5.2 Physical versus Metadata Paths

Because metadata is just information about where things are or who can access them, the physical paths are the actual location of where things are stored.

For example, metadata libraries are data folders that can represent one or more specific folders on the server. The physical path could be c:/ SAS/data/projects or /data/store1/salesproject/forecast. When viewing the data within any SAS BI tool, the end user sees the assigned metadata folder location.

In the SAS Management Console, the SAS administrator defines and maintains the metadata and folder structure. During the initial implementation, this folder structure is defined and implemented.

Refer to Chapter 2, "SAS Enterprise Guide," for suggestions for organizing the folder structure.

The SAS folder structure, such as the one seen in Figure 1.5-1, corresponds to a WebDAV location on the server. WebDAV is an HTML extension that offers security access, version control and other capabilities. Established during configuration and managed by administrators, different versions of SAS BI software have leveraged various WebDAV technologies, including Xythos and SAS Content Server.

The typical report developer views, navigates, and stores content within WebDAV by using only the SAS BI clients discussed in this book. For additional information on leveraging WebDAV for such things as storing batch report content and sharing content that is not SAS content, refer to the SAS Publishing Framework Developer's Guide.

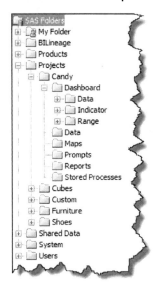

Figure 1.5-1 SAS Folders structure

1.5.3 Libraries

Libraries are data folders. They can represent one or more specific folders on the server that have SAS datasets within them. They can also provide the information SAS requires to access relational database management system (RDBMS) structures, such as Teradata or SQL. Libraries are defined by administrators in SAS Management Console, or with appropriate access they can be defined by SAS Enterprise Guide users.

Chapter 2

SAS Enterprise Guide

Using a Favorite Tool in a New Way

Chapter 2

SAS Enterprise Guide

Using a Favorite Tool in a New Way

SAS Enterprise Guide is a Windows application that provides a point-and-click desktop interface to SAS. It communicates with SAS software to access data, perform analysis, and generate results (such as graphics, PDF files, and HTML files). From SAS Enterprise Guide you can access and analyze many data types, such as SAS data sets, Microsoft Excel spreadsheets, and third-party databases. You can either use SAS tasks or write your own SAS code to perform your analysis. This application helps you produce results easily, regardless of your knowledge of SAS software.

Depending on how your environment is structured, you can use SAS Enterprise Guide to extract, transform, and report on data all from within the same application. Because SAS Enterprise Guide allows you to run code, you can build information maps and OLAP cubes as well as create prompts and stored processes. This chapter focuses on how SAS Enterprise Guide is used with SAS Business Intelligence.

To learn more about SAS Enterprise Guide, see support.sas.com for documentation, training, and other reference material. Also note that the SAS Enterprise Guide online help is an excellent reference.

2.1 Getting Started

Here is a quick view of the tool and some suggestions for what you need to use the examples in this chapter.

2.1.1 Quick Tour

When you open SAS Enterprise Guide, the application window appears.

Figure 2.1-1 SAS Enterprise Guide quick tour

1	The menu bar and toolbars allow you to access the SAS Enterprise Guide functions.
2	The Project Tree pane shows the project that you have open. SAS Enterprise Guide uses projects to manage each collection of related data, tasks, code, and results. With projects, you can run multiple tasks on the same group of data files, and you can save a project to run later or to schedule to run in batch mode.
3	The Resources pane enables you to access the **Task list**, SAS folders, **Server list**, and the Prompt Manager. By default, the Resources pane displays the **Server list**.
4	The Workspace displays your data, code, logs, task results, and process flows.

2.1.2 Prerequisites

To complete the examples in this chapter, you need the following:

- SAS Enterprise Guide installed on your Windows desktop.

- Authorization to access the SAS Metadata Server. The SAS administrator is responsible for granting access and determining what privileges each user is given. Before you can access the SAS Business Intelligence environment, you need to contact the SAS administrator for the server location and the user information. In most situations, you can use your network user ID and password.

2.2 SAS Enterprise Guide on the SAS Intelligence Platform

When you connect to the SAS Intelligence Platform, you are connecting to a metadata environment, which contains definitions of the data you are allowed to access. This central metadata environment operates as a service on a remote server and includes definitions for objects such as users, workspace servers, and libraries.

After connecting to the SAS Metadata Server, one or more SAS Application Servers are available to run your SAS Enterprise Guide project and tasks. The application service waits for and fulfills requests from client applications for data or services. In a typical installation, two SAS Application Servers are available and are labeled as SASMeta and SASApp. SASMeta is recommended for use by administrators in managing the SAS Metadata Server. SAS Enterprise Guide users should select SASApp for all programs.

2.2.1 Connecting to the SAS Metadata Server

To connect to a SAS Metadata Server, you need to create a profile in SAS Enterprise Guide. The profile explains to the SAS Metadata Server who you are and what environment you want to access. While environments vary among organizations, consider the following example setup.

A separate environment is set up for development, test, and production. Each environment has a set of libraries and data defined, based on how the environment is being used. Some organizations might have several different production areas; one is used for the customer service data and the other has manufacturing data. The SAS administrator defines these environments in the SAS Metadata Server, along with who is allowed access to each.

Before connecting to the server, your identity and privileges are verified with the SAS Metadata Server. In the following figure, you can see an example of how your workstation connects to the remote server, where you can access the environment called Production.

Note: You can have multiple profiles defined, but you can connect to only one environment at a time. However, you can open multiple SAS Enterprise Guide instances on your machine to connect to different environments at once.

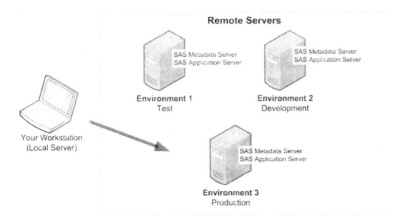

Figure 2.2-1 Connecting to a remote server

2.2.1.1 Viewing the Environment

When you have connected to a SAS Metadata Server environment, the Server List pane displays the libraries and files available to you, as shown in the following figure. Libraries are assigned or unassigned, which is denoted by color. Assigned libraries are identified with a yellow icon and unassigned libraries are identified with a white icon.

 Active server connections are noted by a checkmark on the server icon.

Within each server, there are icons that you can select called **Libraries** and **Files**.

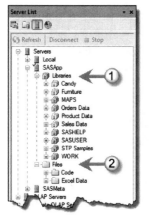

1.   Libraries are shortcut names for known directory locations. Libraries can contain SAS data sets and can point to relational databases, such as Teradata and Oracle. These are created with a LIBNAME statement in SAS code, using the SAS Enterprise Guide wizards or within SAS Management Console.

   Double-click unassigned libraries to assign them for immediate use. You must have authorization to access the data.

2.   Files represent a shortcut into the directory structure on the computer where the SAS server is running. For example, if you want to open a Microsoft Excel file on a server that is defined in your repository, you use the Files node to locate and open the file.

Figure 2.2-2 List of libraries

You can point the file shortcut to another location by changing the SASApp—Workspace Server properties. See Section 2.7.2, "Setting a Default Server-Side File Folder," for more information.

When you connect to the remote environment, SAS assigns a temporary storage folder to you called WORK. This folder is active only while you are connected to the server. Once you disconnect, any information stored in this area is lost.

You can run code on different servers or even locally. See Section 2.6.3, "Choosing Where to Run Your Program."

2.2.1.2 Creating Profiles

Before you can create a profile, you need to know the name or address of the environment and the required user information. Many organizations have configured the environment to use the network user ID and password. Your SAS administrator has specific instructions for completing a profile. You'll notice that if a profile has already been created within SAS Add-In for Microsoft Office, it also appears in the Profiles list viewed from SAS Enterprise Guide.

To create the profile, do the following:

1.   Select **Tools > Options**. The Options window appears.

2.   Select **Administration** from the list. In the Administration panel, click the **Modify** button to manage profiles. The Connections window appears.

     The Connections window shows all of your profiles. When you are connected to an environment, the profile is active and a blue icon is displayed next to it. From the Connections window, you can create, modify, or delete profiles.

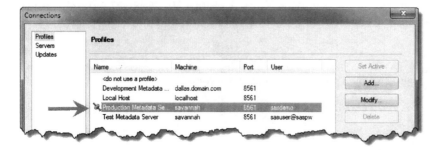

3. Click the **Add** button to start a new profile. The Modify Profile window appears.

    The following figure contains an example profile for a production environment.

    - The **Name** field contains the environment name, which is **Production Metadata Server**. This name is displayed in the **Connections** list.

    - The **Machine** area indicates that this is a remote environment on savannah on port 8561.

    - The **User** and **Password** field values are supplied. This login information is being saved with the profile. When starting SAS Enterprise Guide, the user supplies the password once. SAS Enterprise Guide stores the password and uses it behind the scenes to authenticate the user.

 If the organization security policy does not allow for stored passwords, administrators can disable this functionality during the installation and configuration of the SAS software.

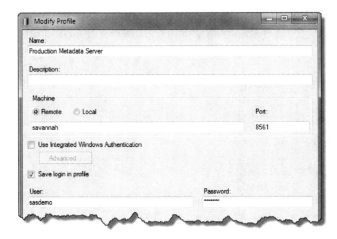

4. After completing the profile, click **Save** to return to the Connections window. The new profile appears in the list. Click the **Set Active** button to start using the environment.

 In the lower right of the main window, you can see which remote environment is active. Click **Connection** to go directly to the Connections window, where you can make modifications to or change the active remote environment.

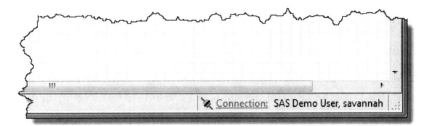

2.2.2 Creating New Metadata Libraries

If the SAS administrator has granted the necessary permissions, you can define libraries within the metadata so other users can also view and use the data. Before you start this process, you need to identify where the data can be stored within the server environment.

To create a new metadata library, do the following:

1. Select **Tools > SAS Enterprise Guide Explorer** from the menu bar to open the SAS Enterprise Guide Explorer window. **Select File > New > Library** from the menu bar to start the New Library wizard. This wizard guides you through the process to create a library.

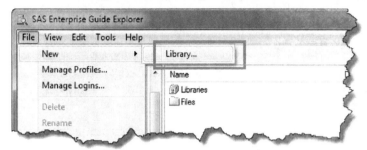

2. In the first step, enter the name of the library and a description of its contents. The name must be unique for each server on which the library is created. In the following figure, CANDY is the common name for the library. Click **Next**.

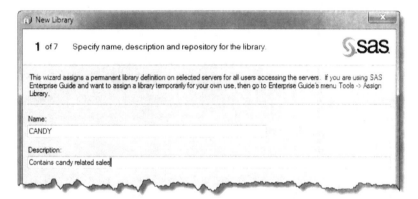

3. The next step determines how the library is assigned.

    a. Specify the LIBREF for the library in uppercase letters. The LIBREF is used in a SAS program when you are defining a library. The LIBREF is a shortcut name for the library. A LIBREF is limited to 8 characters.

    b. You have three choices for assigning the library.

        *   Choose **SAS Enterprise Guide** to specify that SAS Enterprise Guide assign the library using the LIBNAME statement stored within the SAS Metadata server.

        *   Choose **SAS Server** to specify that the library is assigned in an autoexec file or through a METAAUTOINIT option. Use this choice when the library has already been assigned.

        *   Choose **Metadata LIBNAME Engine** to specify that the library is assigned using the metadata LIBNAME engine.

        If you select the **Show only tables with metadata definitions** check box, then only tables that have metadata definitions appear when the library is selected in SAS Enterprise Guide.

The **Allow tables to be created and deleted** option specifies that you can read, update, delete, and create tables that have metadata definitions. You can also insert records, create columns, and change column properties on tables. Tables that you create cannot be read until they have been registered in metadata.

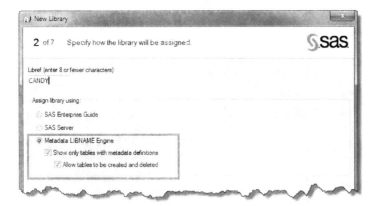

Select one or more servers from the list of available servers. A separate library is created on every server that you select. Click **Next** to continue.

 In most installations, SASApp is the only server instance that SAS Enterprise Guide users can use. SASMeta has limited availability and is primarily used for metadata management tasks by the SAS administrators.

4. You can choose to include a file system, database location, or WebDAV.

    Select the engine type, engine, and the path. The CANDY library contains SAS data sets so the engine type is **File System**. You can also connect directly to a database system, such as Oracle or Teradata.

    Type the path of the physical location of the files or provide the database schema information.

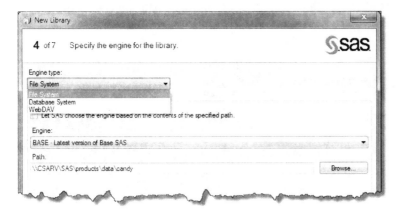

5.   Specify options and option values for the library.

To enter an option name or value, click on a blank table cell and type the name or value.

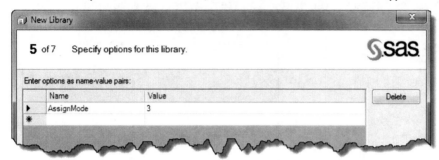

A common option is AssignMode, which defines how the server creates the library and how users can interact with the data tables.

Value	Server	Behavior
0	SAS Enterprise Guide	Created within the SAS Enterprise Guide session and available only to that user during that session.
1	SAS Metadata Server	METAOUT=ALL enforces restrictions to prevent additions and modifications that would cause library metadata to become out of sync with the physical tables.
2	SAS Metadata Server	METAOUT=DATA makes it possible to add, modify, and delete tables within the library.
3	SAS Application Server	Pre-assigned data library that is available to all SAS BI clients.
4	SAS Metadata Server	METAOUT=DATAREG allows users to read, update, and delete tables that are already defined in the metadata.

6.   Use the folder tree to choose the folder that contains the library definition. You can create new folders from the folder tree in SAS Enterprise Guide Explorer or from the folder view in SAS Enterprise Guide. If you do not select a folder, the library is created in the SAS Shared Data folder by default. Click **Next**.

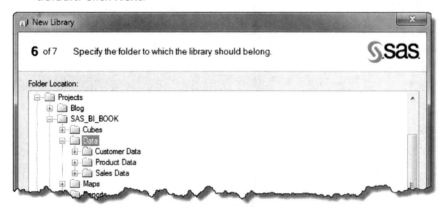

7. Review the information entered in the wizard and click **Finish** to create the library. In the following figure, you can see the library that was created. The Candy library is listed in the SASApp server.

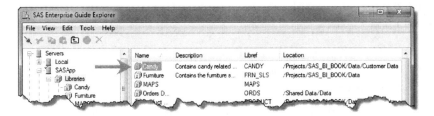

2.2.3 Updating Library Metadata

It is imperative that the metadata and data tables stay in sync with each other. If the data table has columns added, removed, or otherwise modified (including variable lengths, formats, etc.), the other BI clients that use the data tables can have issues when they try to use the data. For example, when the data table and metadata are out of sync because they have different information on what columns exist within the table, SAS Web Report Studio users see a JavaScript web error in the browser.

Use the Update Library Metadata utility to update the metadata from SAS Enterprise Guide. In this example, the Candy library created in the previous example has several data sets that are not registered. When you view the library in SAS Enterprise Guide, the Candy library appears empty, even though there are five data sets in the logical directory. In this example, you will learn to update the metadata library with the data sets.

1. Select **Tools > Update Library Metadata** from the menu bar. The Update Library Metadata window appears.

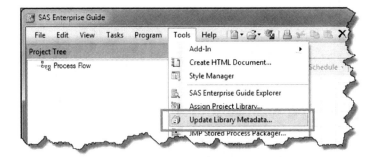

2. Select the server from the drop-down list. The list of available libraries depends on the server that you select. Select a library from the list of available SAS libraries. Click **Next** to run a report or to update the metadata for the selected library.

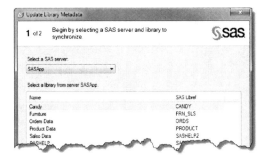

3. Select one of the following options and click **Finish** to run the report or update the metadata.

 When you make changes to metadata library tables in SAS Enterprise Guide, you must manually sync the changes with the metadata.

The first two options return a simple report on the metadata library contents. The last three options make modifications, updates, and deletions to the metadata library contents. These options allow you to make the changes to the tables depending on what you need to do.

Update and add table definitions in metadata repository with the actual tables and columns	Adds and updates the metadata table definitions, including table columns, indexes, and keys, from the physical library to the metadata repository.
Update only the existing table definitions in metadata with the current column information	Updates the table definitions in the metadata repository for those tables in the physical library that have metadata.
	Use this option when the physical library has tables you do not want added to the metadata repository or you have made changes to existing tables.
Delete obsolete entries from the metadata library	Deletes table from the metadata repository when the table no longer exists in the physical library.

For this example, you need to update the new Candy library with the physical library tables, which is option 3. When making changes, you might need to specify a user ID and password for an account that has sufficient privileges.

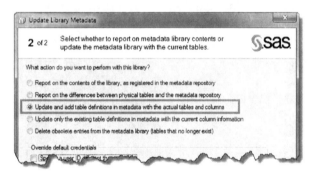

A report is generated that shows the Candy library had six tables added. The report lists the table names and some metadata table information.

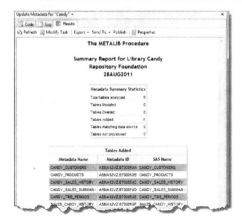

SAS Enterprise Guide creates a task in your project tree for this action. You can rerun the task as needed or add the code from the task to your program.

2.3 Opening SAS Information Maps and OLAP Cubes

Using SAS Enterprise Guide, you can open SAS Information Maps and OLAP cubes to use in your reports.

2.3.1 Opening a SAS Information Map

Information maps provide business users with a user-friendly way to query data and get results. An information map contains data items and filters, which you can use to build queries. You can use information maps from SAS Enterprise Guide, when the map is based on relational data tables. Information Maps built using OLAP cubes are not accessible by SAS Enterprise Guide.

 In Chapter 6, "SAS Information Map Studio," you will learn how to create this SAS Information Map.

The following example demonstrates how to open the information map called Candy Customer Sales.

1. You can access the information map by selecting **File > Open > Information Map** or from the Resources pane. Once you have opened an information map, the Open Information Map – Candy Customer Sales window appears.

2. In the following figure, the available data items are on the left and the selected variables are on the right. The selected variables are the ones you want to import into SAS Enterprise Guide. In the Selected variables area, detailed information about each variable is shown.

   To display predefined computed items, select the **Display aggregated values (group by category)** check box. If you select this option, then the inclusion of these items results in an aggregated data set that is summarized according to the information map definition, which displays the measure data items across the categorical data items.

   Aggregation is the process of grouping data using an operation that produces a statistic such as a sum, average, minimum, or maximum. If this option is cleared, detail data is displayed in the data set. Detail data is factual information that is not summarized (or is partially summarized) that pertains to a single area of interest, such as sales figures, inventory data, or human-resources data.

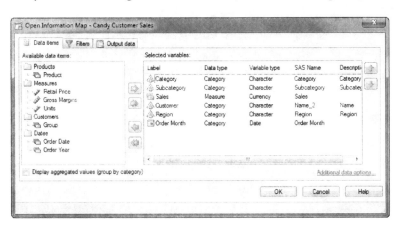

3. Click **OK** to import all data from the information map into the Work library. The result appears in the SAS Enterprise Guide work area. Click the **Code** tab to view the code that imported the map, the log, or the data set.

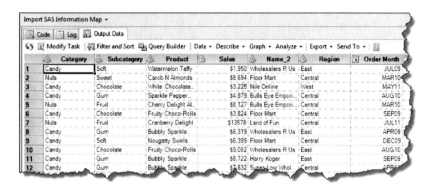

2.3.2 Filtering the Information Map

An information map can represent a large data set. After importing the data, you might realize that you wanted to work with only a certain period and region. You can filter the information map as you import it or after you have imported it.

The following example shows how to modify the task to add a filter:

1. With the information map open in your work area, select **Modify Task** from the menu. The Open Information Map window appears.

   Click on the **Filters** tab to make changes. This information map already has some filters available. Move the filter you want to use to the Selected Filters area and click **OK**. In the following figure, the Past12Months and RegionFilter filters were selected. The RegionFilter filter indicates that it has prompts available, which means you are asked to select a region before the information map generates. The Past12Months filter returns 12 months of data and does not require any user input.

2. Because you selected RegionFilter, which includes a prompt, the region prompt appears, as shown in the following figure.

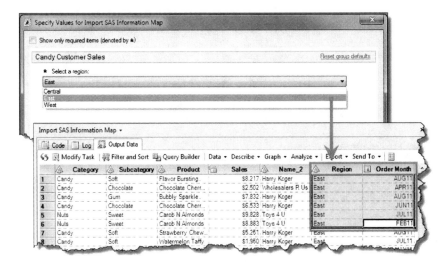

You can use this data just as if it were any other SAS data set. You can join the information map with other maps or create reports from the data.

2.3.3 Working with OLAP Cube Explorer

OLAP Cube Explorer enables you to delve into OLAP cubes, add bookmarks for quick reference, add new measures, and highlight areas for further exploration.

 OLAP Cube Explorer is similar to the OLAP Cube Viewer that is available in the SAS Add-In for Microsoft Office for Excel.

2.3.3.1 Opening an OLAP Cube

You can open an OLAP cube from the **File** menu by selecting **Open > OLAP Cube** and navigating to the folder where the OLAP cube is stored. The OLAP cube opens in the OLAP Cube Viewer window, as shown in the following figure. This example uses the FurnitureSales cube. In Chapter 5, "SAS OLAP Cube Studio," you will learn more about OLAP cubes and how they are created.

The OLAP Cube Viewer opens to a cross-tabulation view and a graph. These items are linked so that when changes are made to one, the other is synchronized and the same changes are displayed. The right pane provides mechanisms to view data, add new measurements, add filters, and create conditional highlights.

 You can right-click on many items for additional choices or to access a favorite task quickly.

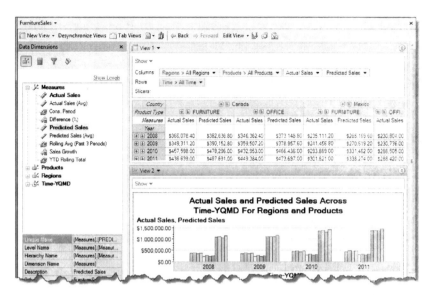

Figure 2.3-1 OLAP Cube Viewer

It is easy to change which data is displayed. If you want to add an item, such as a measure, you can click the item in the Data Dimensions pane and select from the pop-up menu to add it to the column, the row, or as a slicer. You can also select data items already in the data table. In the following figure, the column has the Actual Sales measure and the row contains the Time shown by Year and All Products. If you wanted to view the sales by **Regions** instead of **Products**, click **Products** and then use the **Replace With** choice on the pop-up menu to change to **Regions** easily.

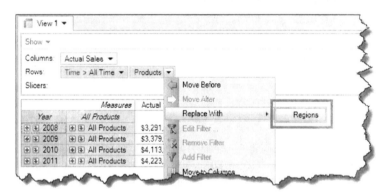

Figure 2.3-2 Change row values quickly from pop-up menu

With some practice, you can easily master the features available to create the views. For a more through explanation of the menu choices, refer to the SAS Online Help.

2.3.3.2 Setting an OLAP Cube Bookmark

As you work with the cubes, you might find that there are particular views that you find yourself creating often. You can create a bookmark that instantly changes the cube to the desired view. Bookmarks are stored within the SAS Enterprise Guide project. To retrieve these for later use, save the project and use it rather than reopening the OLAP cube from the **File > Open** menu. For instance, you might want to create a bookmark to view the same data by region, product, or year. In the following example, you want to have a view by region and product for the current year.

To create bookmarks, do the following:

1. Arrange the columns and rows in the view you want. In the Control pane, select the **Bookmarks** icon to display the Bookmarks pane. Then click the **New Bookmark** icon to add the bookmark.

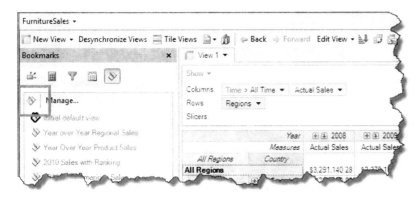

2. In the **Name** field, type a name for the bookmark, and add a description. Click the **Add** button to save the bookmark.

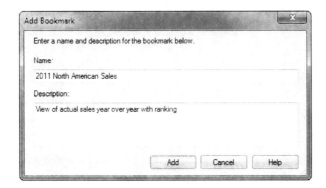

The bookmark is added to the list. Later, if you want to change the bookmark, use the **Manage** button to delete, rename, or reorder the bookmarks.

 To change the initial default cube view, right-click the bookmark and select **Make Default**.

The bookmark is saved within the SAS Enterprise Guide project, not with the cube. To reuse these bookmarks, save the SAS Enterprise Guide project. When you reopen the project , the cube and bookmarks will be available.

2.3.3.3 Creating Custom Measures

As you analyze the data, you might realize that you need more measures. In the following figure, the OLAP cube shows the Predicted and Actual Sales for the past year. To determine how accurate the forecasts were for the year, you need to add a custom measurement that shows the percentage difference.

To create a custom measure, do the following:

1. Click the second icon in the Control pane to display the Customized Items and Sets panel. Then select **New > Calculated Measure**, as shown in the following figure.

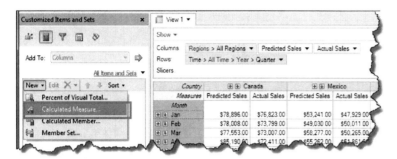

2. The Calculated Measure wizard guides you through the steps. You can create the simple, special analysis, or measures with custom statements. For this example, provide a name and select **Basic Analysis** before continuing to the next window.

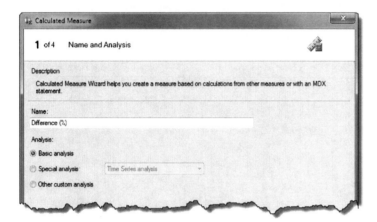

3. In this step, you provide the measures that are actually used to calculate the new measure. In this example, **Percent change** is selected in the **Calculate** field. Then **Actual Sales** and **Predicted Sales** are chosen to complete the calculation.

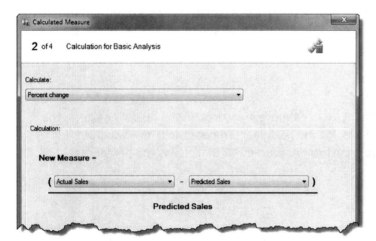

4. For the Other Options window, you need to select the format, who can see the measure, and the solve order.

   - In the **Format** field, there are various formats available, based on your measure. Because this is a percentage value, the Percent format was selected.

   - You can determine if the measure is available only to you or to the general user base. If you set the option to **Publicly**, you cannot change the measure.

 It is a good practice to keep the measure locally at first. After you have tested the measure, you can edit the measure to make it available to others.

   - You do not need to make changes to the **Solve Order** field unless your measure is based on another calculated measure.

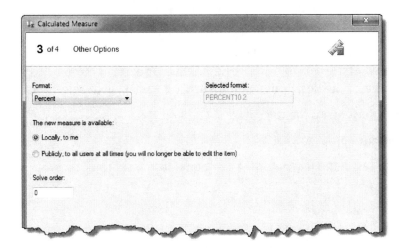

5. The remaining window allows you to review your changes. If the measure is correct, then click **Finish** to create the measure.

   The custom measure is added to your current view. In the following figure, you can see Difference (%) was added after the Actual Sales measure. The measure is also listed with the other measures in the pane on the right.

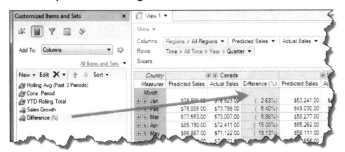

2.3.3.4 Adding Conditional Highlighting

With the conditional highlighting feature, you can set rules for the data display to see any trends or patterns that merit more attention. Conditional highlighting allows you to change the background color, add an icon to the table cell, or change the font. In the last topic, you created a new variable that shows the percentage difference between two measures. Using the conditional highlighting, you can turn the cell background to a different color if the difference in the values is more than 1%.

Do the following to add conditional highlighting to your displayed view.

1. Click the fourth icon in the Control Panel to display the Conditional Highlights panel. Then click the **Add** icon, as shown in the following figure.

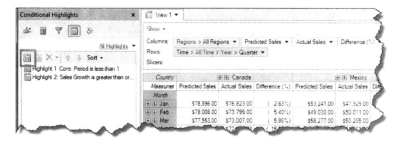

2. To add the conditional highlight, you first create the rule and then how you want it highlighted. In the **Name** field, you can change the name of the rule to be more descriptive. You can edit the **Description** field, but the application creates the description automatically based on your selections.

    To create a rule for the Difference (%) measure, do the following:

    a. Select **Difference (%)** from the **Measure** drop-down list.

    b. Click the **By range** radio button and select the **Is greater than or equal to** choice from the drop-down list.

    c. Type 0.01 in the field.

 In the Highlight panel, you can change the font color, background color, or add an image. By default, any rule changes the background color. However, you can choose to add an icon.

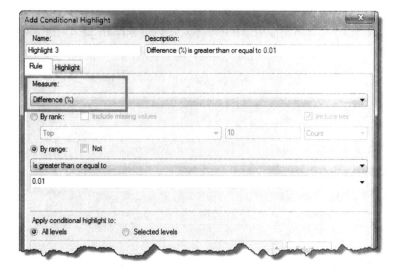

3. Click **OK** to exit. The conditional highlight is added to your current view.

   In the following figure, you can see the highlight is listed with the other measures in side panel. The Difference (%) column has a different background color and a checkmark to the left of all figures. This highlight quickly reveals that few areas exceeded the predicted sales.

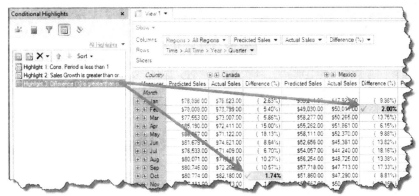

As your skills progress, you will find other OLAP features helpful. You can learn more about these features from the SAS online documentation.

2.4 Using Project Prompting

A prompt is a way to ask the user a question. For example, "Which region do you want to see" or "Which months do you want to use." These questions, or prompts, are asked before a report is generated or data query is created. You can create prompts to accept text strings, numbers (with range validation), single or multiple values from a predefined list of values, date or date-time values, and even variable names for use within SAS task roles. For detailed information about creating and using prompts, refer to Chapter 4, "The Prompting Framework."

If you are familiar with programming in the SAS language, the prompts create macro variable to pass user-specified variables to the program.

2.4.1 Viewing Prompts

To view the prompts, open the Prompt Manager window by selecting it from the View menu. All existing prompts are listed within this window, as seen in the following figure. If any task is using the prompt, you can see the name in the Used By column.

In the following figure, the project has three prompts available: ProductPrompt, RegionPrompt, and YearPrompt. Two of the prompts are already used by two different processes. Those processes have a question mark with the icon to indicate that the prompt is used. You can add, edit, or delete prompts using the Prompt Manager.

In the following example, you will learn how the ProductPrompt was added to the Get Product task.

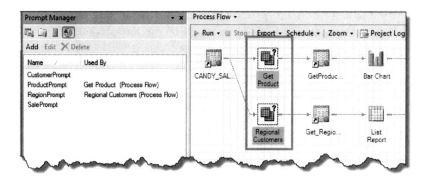

Figure 2.4-1 Prompt Manager area in SAS Enterprise Guide

2.4.2 Applying a Prompt to Your Query

You can easily add a prompt to a SAS program or task. In the following example, the ProductPrompt is added to the Get Product task. Get Product is based on the CANDY_SALES_SUMMARY data set and uses the Query Builder task.

To associate the prompt, do the following:

1. Right-click **CANDY_SALES_SUMMARY** and select Query Builder from the pop-up menu. Add the fields you want in the data set.

2. Click the **Filter** tab and select **Product** as the filter item. The ProductPrompt is based on the **Product** field in the CANDY_SALES_SUMMARY data set. The prompt allows the user to select one or more products. You need to know this information as you create your filter.

    To add the filter with the prompt, do the following:

    a. Select the operator, based on how the prompt is set up. Because this prompt allows more than one value, select **In a list** from the drop-down list.

    b. Click the check box **Generate filter for a prompt value** to ensure that the filter was created correctly. This will add a macro variable to the code.

    c. In the **Value** field, click the down arrow. Then select the desired prompt from the **Prompts** tab.

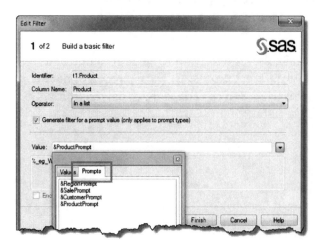

3. Click **Next** to continue. In this window you can verify the prompt information. A special global macro called %_eg_WhereParam was added.

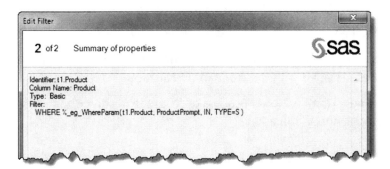

4. Complete the Query Builder and run the task to test the process.

   In the following figure, you can see an example of a prompt that can select more than one value or add a new value.

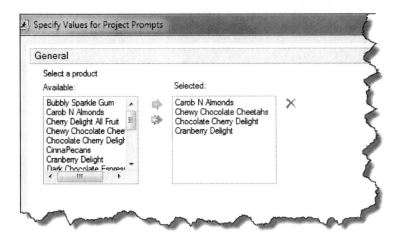

2.5 Using SAS Stored Processes

You can create and run stored processes from SAS Enterprise Guide. Any task that can be done from SAS Enterprise Guide can be made into a stored process. Stored processes provide a method to run a task repeatedly with different values. SAS Stored Processes can be shared with SAS Web Report Studio, SAS Add-In for Microsoft Office, or made available from an Web browser. When a stored process is run from a website, users do not need any SAS software installed on their computer to run or view the report.

Weekly reporting needs are common situations where a stored process is implemented. For example, you might create a report that details weekly product sales by region. This report shows the data from several viewpoints, perhaps a bar chart, a line plot, and some summary data. Your manager and others in your department ask for the report frequently. Each person who requests the report wants to see it for a specific region or product line and for the most recent sales data.

You can create this report once and turn it into a stored process. You can make this report available in different formats, perhaps as PDF or an HTML. Then the user can generate this report on demand, by region or product.

You can also create stored processes for parts of reports that you create frequently. For detailed information about creating SAS Stored Processes, refer to Chapter 3, "SAS Stored Processes."

2.5.1 Running a SAS Stored Process

The following example demonstrates how to run a sample stored process called Shoe Sales by Region. This is one of the SAS sample stored processes provided with the standard installation.

1. You can access the stored process from **File > Open > Stored Process** or from the Resources pane. In the following figure, you can see the sample stored processes.

2. Right-click the stored process name and select **Add to Project** from the pop-up menu, or double-click the name to add it to your active project. The stored process appears in the **Project Flow** and in the **Project Tree**.

3. To run the stored process, right-click it in the Project Tree or Project Flow area and select **Run Sample: Shoe Sales by Region**.

4. When the stored process starts, you are prompted to select from the values that were predetermined by the stored process author.

5. In this example, you are asked to select an ODS style, and you are given the option to display the SAS log. While these prompts are more general, you could also prompt a user to select a region, product, or even the output format.

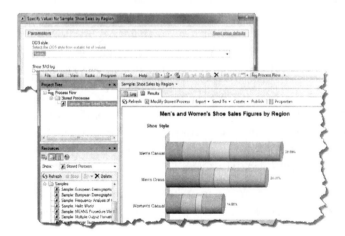

2.5.2 Working with Stored Processes

Typically, a stored process generates results as a SAS Report and uses the default SAS Enterprise Guide style. To override the result format, styles and behavior, use the Properties window.

To change the result format, use the following example:

1. Right-click the stored process name in the Project Tree or in the Process Flow and select **Properties**.

   The **Use preferences from Tools > Options** option shows the default settings for this stored process. The other results options are disabled if this option is selected. For more information about creating your own style or changing the default style, see Section 2.6.1, "Customizing Styles."

2. In the Result Format area, select the report format and the style you want to use.

   There are five choices: SAS Report, HTML, PDF, RTF, or Graph Format. A suggested style is supplied for some of the result formats. For instance, the PDF result format generally produces a nice-looking report with the printer style. However, this recommendation should not prevent you from trying the various styles to find the one that works best for your report. Click **OK** after you have made your choice.

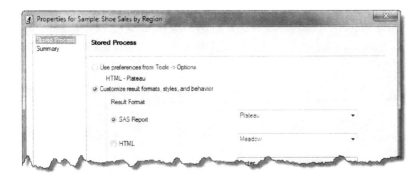

3. You can view the report with the new option by right-clicking the stored process name and selecting **Run <stored process name>**, where *stored process name* is the name of the stored process.

2.6 Tips and Tricks

SAS Enterprise Guide allows you to control publishing to channels or change the look of your report.

2.6.1 Customizing Styles

In SAS Enterprise Guide, a style is used for the report output. You can think of a style as a template that controls the font size and color for headings, tables, and graphs. SAS Enterprise Guide ships with over 40 different styles available for immediate use. These styles create the extra spice in your reports. As you will learn in this section, you can also create your own styles.

2.6.1.1 Viewing the Available Styles

You can view the available styles by selecting **Tools > Styles Manager**. The Style Manager window appears (as shown in the following figure). From here you can review all the available styles. The default style for all reports has a bolded name and is displayed first.

Click a style name in the **Style list** to see a preview of the style. If you find a style you would like to use for all of your report, click the **Set as Default** button. The next time you generate a report, the newly selected style is used.

Figure 2.6-1 Set the default style in the Style Manager window

Some of the styles make better use of space than others. For instance, Plateau, Seaside, Meadow, and Journal produce more compact reports. When you need to display a lot of text, these styles work nicely. A style that has almost no formatting is called Minimal. You might want to use this as a template for building other styles.

2.6.1.2 Modifying Styles

You can change the look of your HTML or SAS Report output by modifying and creating styles. Styles are based on cascading style sheets (CSS) and are a set of specifications that control what your HTML or SAS Report output looks like.

You can change the default fonts, control spacing, add images, or simply change the colors. In the following example, you will learn how to modify a report so the headings are larger, and how to add an image to the top of the report.

To modify a style, do the following:

1. From the Style Manager window, select the **Normal** style and click the **Create a Copy** button.

2. In the Save Style As window, type `Normal_2.css` in the **File name** field and click **Save**.

   Normal_2 appears in the **Style list** with a different icon and a location of **My Style**. Now you can easily identify styles you have created.

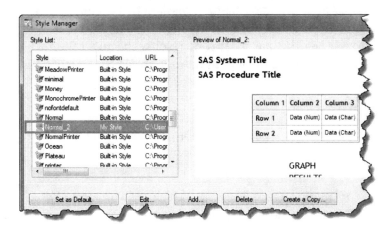

3. Select **Normal_2** and click the **Edit** button to display the Style Editor window.

4. To change the heading, make the following choices in the Text panel:

   a. Click the words SAS System Title in the Elements area or select it from the **Selected element** list.

   b. In the **Text size** drop-down list, select 18pt.

   c. In the **Text style** drop-down list, select Bold Italic.

   d. In the **Horizontal alignment** drop-down list, select Left.

   You can apply these changes to other styles by using the **Apply to Other Elements** button. If you make a change you do not like, click the **Undo Apply** button to return those styles to their original setting.

5. You can use GIF or JPG formatted images for the banner or background image. The banner image is displayed across the top of the page and the background image is displayed on the entire page.

   To add a banner image, make the following changes in the Images pane:

   a. Select the **Use a banner image** check box.

   b. Click **Select** to specify the image that you want to use. In the Open Image window, double-click on the image file that you want to use.

   c. Click **Open**. The image that you select appears as a banner in the preview window.

   Notes:

   - If your report is shared with others, the image needs to be in a location that others have permission to see.
   - You can also type an URL of an image that is available on the web.
   - The default banner size is 471x72 pixels. Your image should be sized to fit this area.

6. Click **OK** to save the changes and return to the Style Manager.

The style is available only for your local SAS Enterprise Guide installation.

2.6.2 Publishing Data and Results

SAS Enterprise Guide enables you to publish data and task results to predefined channels, which function as repositories to which users subscribe. Any information that is published to a channel is delivered to all of that channel's subscribers through e-mail, the SAS Information Delivery Portal, or some other method. You must be connected to a server to publish data and results. The publication channels are defined on the SAS Metadata Server that you connect to using your profile. You might need to change your profile to publish content.

There are two types of subscribers to these channels.

- A content subscriber is a subscriber who is configured to receive packages, which is a bundle of one or more information entities such as SAS data sets, SAS catalogs, or almost any other type of digital content.

- An event subscriber is a subscriber who is configured to receive events, which are well-formed XML documents that can be published to an HTTP server, a message queue, or a channel that has event subscribers defined for it.

The SAS administrator must establish the publication channels and add the subscribers. Refer to Section 2.7.3, "Establishing a Publication Channel," for more information about setting up the channel on the SAS Metadata Server.

To publish content to a channel, do the following:

1. To publish a SAS Report, right-click the SAS Report that you want to publish in the Project Tree or in the Process Flow, and select **Publish**. The Publish to the Enterprise wizard opens.

2. Enter a name and description in the fields. If you want the report to be available for a limited time only, then specify an expiration date, as shown in the following figure.

3. Specify the server where this package is stored. Click **Next**.

4. Specify the channel you want to use. Click **Next**.

5. Ensure that you have selected your report. Click **Finish** to publish the report. You can add other documents to the publishing package using the **Add Item** button. For instance, you might want to include additional documents that contain relevant or supporting information, as shown in the following figure. An Excel file and a PDF file have been included in the package.

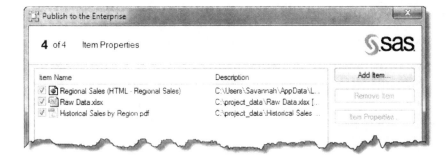

After the package is published to the channel, it can be accessed from the SAS Information Delivery Portal from the Collection portlets. The following figure shows how the package looks from within the portlet. When the user clicks on the package, a new SAS Portal page is opened and the user can see all of the items.

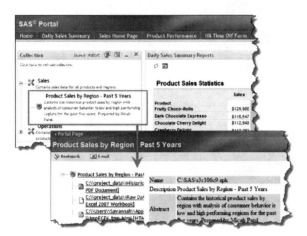

2.6.3 Choosing Where to Run Your Program

When you have connected to the remote environment, any new projects you create automatically use that environment to run the programs. If your installation of SAS Enterprise Guide has been set up to access SAS software on more than one server, then you can easily change where code runs.

Sometimes when the server is busy or you are working with smaller data sets, it might be quicker to run the tasks locally. However, when running locally, you have access only to the data on your local machine.

To change the server and run your code, do the following:

1.   Select **Code > Select Server** from the menu bar.

2.   In the Select Server window, click the server you want to use and click **OK**.

3.   To run your code, select **Code > Run program-name On server-name**. In the following figure, the program is assigned to SASApp, but you could also assign it to the local server by clicking on it. Remember that SASMeta is recommended for use by administrators in managing the SAS Metadata Server. SAS Enterprise Guide users should select SASApp for all programs.

If you want to use another server later, repeat this process and select that server.

2.7 SAS Administrator Tasks

Using SAS Management Console, the SAS administrator can set responsibilities and make system-wide changes that assist all SAS Enterprise Guide users.

2.7.1 Roles and Responsibilities

Within SAS Management Console, SAS administrators grant access to the servers based on the following four roles. Each of these roles provides various user capabilities and offers a framework for distributing access based on user type.

Role	Capability Overview
Advanced	Provides all capabilities in SAS Enterprise Guide.
OLAP	Allows user to view OLAP cubes.
Analysis	Provides basic data analysis, reporting, and other capabilities.
Programming	Provides SAS programming, stored process authoring, and other capabilities.

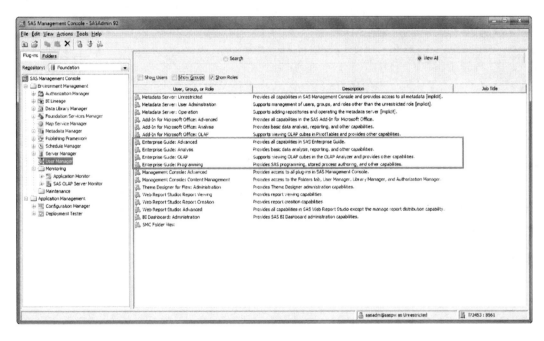

Figure 2.7-1 SAS Enterprise Guide roles from SAS Management Console

If any role requires modification, the SAS administrator should create a new role and select appropriate items on the Capabilities tab found within the Roles Properties window, as seen in the following figure. If you are adding capabilities, it is often easier to add the original role to the **Contributing Roles** tab.

Figure 2.7-2 Capabilities tab for Enterprise Guide: OLAP Properties

2.7.2 Setting a Default Server-Side File Folder

Server files can be accessed directly through the **Files** folder, which is viewable from the Server List window. SAS Enterprise Guide users can quickly access and store files on unfamiliar platforms, such as a Windows user accessing a UNIX platform.

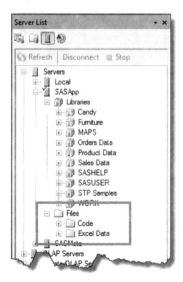

Figure 2.7-3 Files from the SAS Enterprise Guide Server List

The SAS administrator can modify where the **Files** folder points on the server environment by modifying the Workspace Server properties in SAS Management Console and restarting the object spawner.

1. In SAS Management Console, choose **Properties** on the SASApp – Workspace Server.

2. On the **Options** tab, click the **Advanced Options** button and choose one of the following options:

Type	Definition	Example
SAS User Root	Uses the SAS user root directory (or –SASUSER)	C:\Documents and Settings\anhall\My Documents\My SAS Files\9.2
System Root	Uses the system root directory.	\
Path	Uses the specified location defined within this field.	C:\temp

3. Save your changes and restart the SAS object spawner service.

2.7.3 Establishing a Publication Channel

  To add channels or subscribers, the WriteMemberMetadata permission is required on the relevant parent folder.

You can create SAS publication channels based on topics (3rd Quarter Results), organization (Business Unit X), user audience (new product release), or any other category. Once defined, authorized users can subscribe to the channels and automatically receive information whenever it is published.

Note: To publish to a channel, the publisher must have Write permission to the channel. You can modify an existing channel to change the permissions for a user through the **Authorization** tab.

This following example shows how to create a Sales channel to distribute sales results for each product and each region. This channel distributes information to the sales organization, regional management teams, and others in the management team.

1. Open SAS Management Console. On the **Plug-ins** tab, navigate to the **Environment Management > Publishing Framework** area.

2. Right-click **Channels** and select **New Channel** from the pop-up menu, as shown in the following figure.

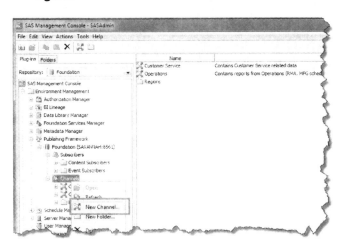

3. In the next window, provide the channel name and other information that helps the users understand the contents or find it in a search.

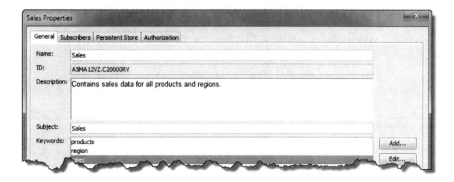

4. Click the **Subscribers** tab. There are two tabs in this window: **Content** and **Event**. There are two types of subscribers to these channels.

    Add the subscribers to the desired method. In this example, the subscribers are added to the **Content** tab so they receive the information when a new package is available.

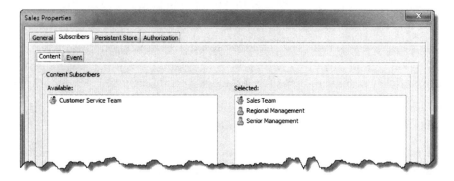

5. When information is published, it should be retained. The following example uses a WebDAV location to store content. WebDAV is a file folder-structured location accessible from the web that maintains file- and folder-level security.

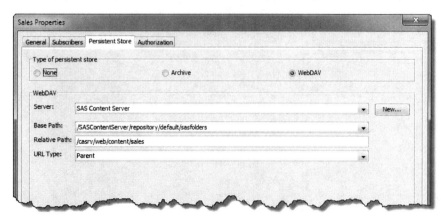

6. The next window shows the results and allows you to make changes. Click **Finish** to establish the channel. In the following figure, you can see the **Sales** channel is now available.

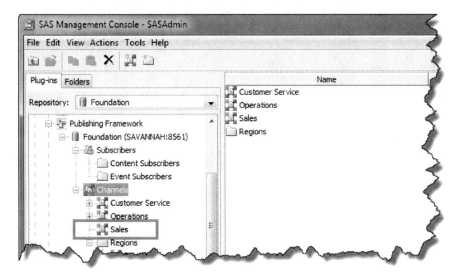

Chapter 3

SAS Stored Processes

Data and Reports—On Demand!

Chapter 3

SAS Stored Processes

Data and Reports—On Demand!

SAS Stored Processes are SAS programs stored in central location that can be run on demand. Stored processes can perform a variety of tasks, such as generating data sets, creating complex reports, and even building Web applications. Almost anything you can do in SAS programs, you can do with stored processes.

Because stored processes are kept in a central location and shared by many applications, it helps organizations consolidate programming silos into one, reducing duplication and improving efficiencies. You can access a stored process from the SAS BI clients (SAS Add-In for Microsoft Office, SAS Web Report Studio), from the Web applications (SAS BI Dashboard, SAS Information Delivery Portal), and from a Web browser. When you run a stored process from any of these locations, you are accessing consistent results. Because the stored process is run on demand, it can use the data that was made available most recently.

End users do not need to have SAS software installed on their computer or device to run the stored process or view the result; they only need permission to access the information. This provides additional security for your data, because the applications that access the data are stored in a central, secure location. Additionally, with the SAS program in a central location, you can make modifications to the code that flow through the entire system, allowing you to ensure that everyone is using a consistent report.

This chapter explains how to create a simple SAS program and convert it into a stored process with some complex functionality. While SAS programming experience is not required, the chapter assumes the reader has a minimum level of SAS programming skills. You can create stored processes using tasks in SAS Enterprise Guide, but without programming knowledge improving or extending the stored process capabilities can be difficult.

3.1 Getting Started

This topic helps you locate the stored process home page and understand the basic requirements needed to start creating SAS Stored Processes.

3.1.1 Quick Tour

SAS Intelligence Platform provides a Web page that lists the registered stored processes. Using a Web browser, go to the SAS Stored Process home page, which is in a location similar to the following address. Note that this address is case sensitive.

```
http://server name:port number/SASStoredProcess/do
```

The following figure shows the default stored process home page. Your organization might have already made changes to the page, so it might appear differently. The stored process samples are available as links.

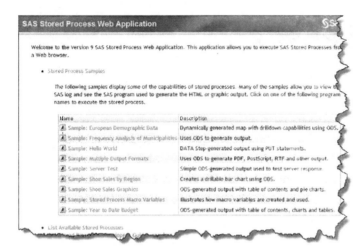

Figure 3.1-1 Quick tour

3.1.2 Prerequisites

To follow along with the examples in this chapter, you need the following:

- Your SAS administrator must provide proper permissions in the metadata and access to appropriate folders to save the stored process.

- SAS Enterprise Guide access to create and register the stored process.

- Basic understanding of the SAS programming language.

3.2 Running SAS Stored Processes

To run a stored process from the Web browser, navigate to the Web location and click the name of the stored process. In the following figure, a stored process called Query Regional Sales was selected from the left pane. The Query Regional Sales stored process appears in the right pane.

This is a straightforward stored process that prompts the user for the region, minimum sales amount, and a time period. The resulting report, Customer Sales by Region, appears in a new browser window. As requested in the prompts, the report shows the all sales greater than $6500 in the Central Region for the previous month. In the next example, you will learn how to create this stored process.

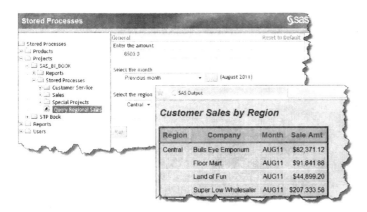

Figure 3.2-1 Example Query Regional Sales stored process

3.3 Creating SAS Stored Processes

A stored process consists of a SAS program file and is set to run on a specific SAS logical server. This information is defined within the SAS Metadata Server to allow the stored process to work. A stored process can be registered on only one logical server; however, the same SAS program file can be used in multiple stored process definitions.

There are three steps to creating a stored process: create a SAS program, choose the server, and chose the output method. This section provides a detailed discussion about how to create stored processes. If you want to create a stored process without this behind-the-scenes knowledge, refer to Section 3.4, "Creating Your First Stored Process."

3.3.1 Create a SAS Program

You can create a SAS program or a task that can be used repeatedly. In the preceding example, the stored process created a simple report about regional sales for a selected time period.

The following program contains the SAS code that was used to create the stored process. The code has three input parameters that are used in the WHERE statement. The input parameters are defined by assigned values to macro variables with %LET statements. Before the code is executed, the SAS engine substitutes the macro variable with the assigned value. So &sale. becomes 5000, &region. becomes East, and &time_period becomes 01JUL2011.

Also, notice that the character macro variables use double quotation marks. This ensures that the variable is properly passed and decoded. The date variable is passed as a character variable, so it uses quotes and is appended with the letter "d" for SAS to interpret into a standard SAS date field.

In Section 3.6.1, "Understanding Macro Language Fundamentals," there is more information about using macro variables.

```
/*=======================================================*/
/*Macro variables used to query the dataset */

 %let sale = 5000;
 %let region = East;
 %time_period=01JUL2011;

libname mylib meta library="Candy";

title "Customer Sales by Region";
proc report data=mylib.sales_candy_history nowd;

 WHERE sale_amount gt &sale.  /*Numeric variable*/
               and salemonth = "&time_period."d /*Date variable*/
               and region in ( "&REGION." )  /*character variable*/
;
column region company salemonth sale_amount;
       define region/group;
       define company/group ;
       define salemonth/group format=monyy.;
       define sale_amount/format=dollar12.2;
run;
/*=======================================================*/
```

Program 3.3-1 SAS code with %Let statements

3.3.2 Add the Stored Process Codes

To change the code to a stored process, you need to add the stored process special codes:

*ProcessBody   This is a required comment. It indicates to SAS that stored process code is ahead.

%STPBEGIN   This macro variable generates the initial Output Delivery System (ODS) statements to publish content to the client.

%STPEND   This macro variable closes the ODS destination.

 The SAS Stored Process wizard in SAS Enterprise Guide includes these special codes by default within the stored process.

In the following program, you can see how a simple SAS program looks after applying the stored process codes. The %LET statements were changed to global macro variables. A global macro variable simply means that these macro variables are available for all parts of the stored process to use.

In this case, prompt values are returned to the stored process as macro variables. The %global statement ensures that the value is available for the entire stored process to use. Later, in Section 3.6.1, "Understanding Macro Language Fundamentals," macro variables are discussed in detail. Right now, you just need to understand that a global macro variable is how the stored process works with the prompts to collect input information.

```
/*=== Start the Stored Process ============*/
*ProcessBody;

%global sale region time_period;

libname mylib meta library="Candy";

%STPBEGIN;
title "Customer Sales by Region";
proc report data=mylib.sales_candy_history nowd;
WHERE sale_amount gt &sale.
              and salemonth = "&time_period."d
              and region in ( "&REGION." )
;
column region company salemonth sale_amount;
      define region/group;
      define company/group ;
      define salemonth/group format=monyy.;
      define sale_amount/format=dollar12.2;
run;
%STPEND;

/*=== End the Stored Process ============*/
```

Program 3.3-2 Sample stored process

3.3.3 Choose the Appropriate Server

While it is natural to think of a server as hardware or a physical location, when you are learning about these servers, it might be easier to think of these servers as applications that perform actions. There are two logical servers available for use.

Stored Process Server This server completes requests under one user identity, SASSRV. A group of sessions is always available and waiting for a user to start a stored process. Each time a stored process is run, a session is assigned to that process. If the user reruns the stored process, a different session is used.

The server distributes or balances the requests across the sessions, thereby providing high performance and allowing for scalability.

Workplace Server This server assigns a single session to each stored process and each user. Therefore, this server has an increased security advantage over the Stored Process Server.

Output can be provided only as a package or through a component such as SAS Web Report Studio.

Use the following table to determine the server that best meets your stored process code requirements.

Task	Logical SAS Stored Process Server	Logical SAS Workspace Server
Run a stored process from a Web browser and output results to the Web browser	X	
Run a stored process within an information map		X
Use sessions to reduce query processing for related stored processes	X	
Access a stored process from SAS Web Report Studio, SAS Add-In for Microsoft Office, or SAS Enterprise Guide	X	X

Table 3.3-1 Overview of SAS Stored Process Servers

The following figure is a quick reference for the features of each logical server.

SAS Stored Process Server	SAS Workspace Server
☑ Multi-user server ☑ Uses single, shared identity for all requests ☑ Uses streaming output, allowing Web access to results ☑ Supports sessions	☑ Single-user processing ☑ Higher security with single identity ☑ Uses package output ☑ Ability to access data ☑ Execute client-submitted SAS code ☑ Supports SAS Information Map use

Figure 3.3-1 Quick reference for logical servers

3.3.4 Register the Stored Process Metadata

When you register a stored process, you need to provide a name, where the code is located, what parameters are used, and how the results or output should be treated. This is how the SAS Metadata Server knows the stored process exists.

The SAS administrator must define a place for your stored process code on the server. Refer to Section 3.8.2, "Setting Up a Source Code Repository," for information on how to set up these physical paths.

3.3.5 Choose an Output Device

After the stored process is complete, the result is displayed. The result might be a Web page, a PDF or RTF file, or in some cases, a data set. Based on the output, you can use one of two different devices: streaming or package.

Streaming Output This output is a written as a data stream to the client, such as an HTML or XML file. These results are immediately available to view in a Web page.

The SAS Stored Process Server uses this option.

Package Output This output is written to a package that can be viewed immediately or later. A package can be any combination of data sets, images, and files. Both servers can use this method.

There are two kinds of packages: transient and permanent.

- Transient output is available when the client is connected to the server. This is a good way to deliver text and graphics output. Content is dynamically removed from the storage device after it is no longer in use.

- Permanent output is stored in a location, such as the WebDAV repository or a server file system. This output can be used immediately or made assessable at any time. This output can be published to a channel or e-mailed to users. You can also have some security on the file, such as assigning a user name and password.

3.4 Creating Your First Stored Process

You can use SAS Management Console or SAS Enterprise Guide to register your stored process. The advantage to using SAS Enterprise Guide is that you can develop, test, and create the stored process within the same application. This topic explains how to use the SAS Stored Process wizard in SAS Enterprise Guide. For more information about SAS Enterprise Guide, refer to Chapter 2, "SAS Enterprise Guide."

For this example, you will create the stored process shown in Figure 3.2-1, based on the code in Program 3.3-2. This report queries the user for the region, month, and order amount to display in a simple report. This stored process uses three prompts that have already been created. If you want to learn more about creating a prompt, refer to Chapter 4, "The Prompting Framework."

Using the SAS Stored Process wizard, you can quickly register the stored process using the following steps:

1. From SAS Enterprise Guide, you should already have created a program to use. Right-click the program icon and select **Create Stored Process**. The Create New SAS Stored Process Wizard window appears.

2.   In the Create New SAS Stored Process Wizard window, you start the first of six steps. The first step is to name and describe the stored process.

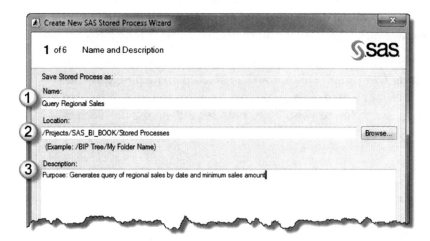

	Field	Description
1	**Name**	Text used by the metadata. This text is displayed to the end users.
		Select a name that describes what the process does so the users can quickly find their stored process. For example, `Get Regional Furniture Sales by Month and Year` is better than `Furniture Sales`. When your list of stored processes starts growing, even you might have trouble recalling the differences between them.
2	**Location**	Use the **Browse** button to navigate to the location where you want to save the stored process on the SAS Metadata Server.
3	**Description**	All the remaining fields are optional. However, as you create more stored processes, it is useful to provide additional details that assist with locating the stored process, remembering the requestor, and determining who is responsible for maintaining it.

3.   Because you selected the Create New SAS Stored Process wizard from an existing code node, your SAS code is loaded automatically. You can edit the code in this window. Because you started the stored process from the code, the link between the code and the stored process is maintained. You might find that editing the code outside of the wizard provides more editing features that assist with maintaining the code.

   Syncing code from SAS Enterprise Guide to the SAS Stored Process provides a better coding interface. To keep the links enabled between the initial code node and the stored process, save and use this SAS Enterprise Guide project to make changes in the future.

By default, the SAS Stored Process wizard adds all of the following codes:

Stored Process Macros	Adds the %STPBEGIN and %STPEND macro variables.
Global Macro Variables	Adds the %GLOBAL for any macro variables in your code.
LIBNAME References	Adds the LIBNAME statements to ensure that SAS understand where the source data is located. Use data in a library that is available to others to ensure that they can access and see your results.

In some cases, you might want to control what is used in the stored process code. On the **Include code for** button, you can select which code is added. As you explore the examples in this chapter, you will discover some instances where the codes are manually added.

For this simple code, the LIBNAME statement was kept, but all other special coding was removed. The SAS Stored Process wizard adds the necessary codes.

 Click the notepad icon in the bottom left corner to see how the wizard modifies your code.

4. In this step, you define where you want the stored process to run, where the code is located, and how the output is created. Do the following to complete this window:

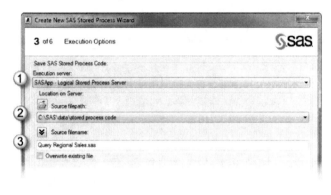

	Field	Description
1	**Execution server**	Select the server that executes the stored process. Refer to Section 3.3.3, "Choosing the Appropriate Server," for more information about the server types.
2	**Source filepath**	Select where the SAS code is stored. This is the physical program location. Click the **Folder** button to navigate to the area your SAS administrator has made available for storing code.
3	**Source filename**	Type a program name. SAS code is stored in a source code repository (listed in the **Source filepath** field). When you save the SAS code, use the .SAS extension.
4	**SAS Result Type**	Select the result type that you want. Refer to Section 3.3.5, "Choosing an Output Device," for more details about the types.

 If your stored process creates only a data set and there is nothing for the user to view, you can leave both check boxes empty.

5. In the next window, you can assign the prompts to your stored process.

    You can add as many prompts as you need. For this example, you must have a prompt for every macro variable in the WHERE statement. Otherwise, the stored process generates an error and does not run.

    There are three different methods for using prompts with stored processes: using a project prompt, creating a prompt, and using a shared prompt. You can use one or more of these methods within the stored process. More information is available in Chapter 4, "The Prompting Framework."

- If you created the project prompts in SAS Enterprise Guide, those same prompts might be converted into the stored process. Select **New > Project prompt copy** to add the project prompt. In the following figure, the Create New SAS Stored Process Wizard window lists the project prompt time_period as a choice.

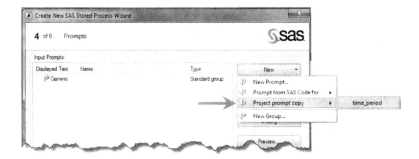

- If you want to use a shared prompt, select **Sharing > Add Shared**. Navigate to the area where the shared prompts are stored and select the desired prompt. In this example, region is available as a shared prompt.

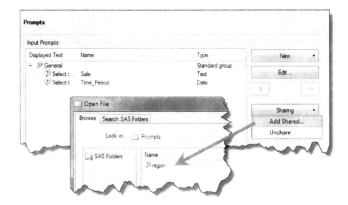

The code is currently set up to filter on only one selection of region; therefore, this shared prompt would not provide the expected results, as it would query on only one of the user's selected values. Refer to Section 3.6.1.5, "Integrating Multiple Selection Prompts into a Stored Process," for information on coding filters for multiple selection prompts.

- If you want to create the prompt, select **New > Prompt from SAS Code for <macro variable>**. You are guided through the Prompt wizard to create the prompt. All macro variables in the code are listed.

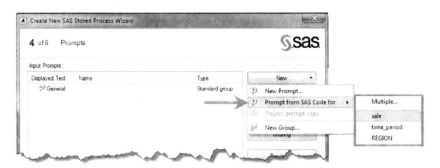

The following table provides the details on what prompt values and options to select if you are creating these three prompts for this example.

Prompt Name	Displayed Text	Prompt Type	Method for populating prompt	Default Value
Time_period	Select the month	Date	User enters value	Previous month
Region	Select the region	Text	Static List	Central
Sale	Enter the amount	Number	User enters value	6500

In the following figure, you can see that all of the prompts were added. The **Name** column contains the prompts, which match the macro variables used in the WHERE statements in the code.

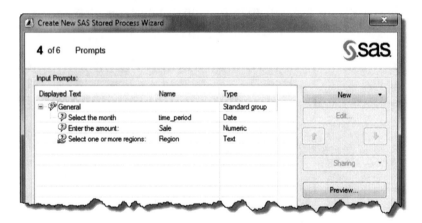

6. In the next window, you can specify additional data sources or data targets that you can define. In some cases, the stored process needs to accept a range of cells within Microsoft Excel. By using these additional data source options within the stored process, code and prompts can offer significant value and usability.

7. In the last window, you can review the stored process to ensure that everything is set up correctly. In the following figure, the changes to the initial SAS code are displayed. When you scroll through the SAS code, you can see that the global macro variable, and the %STPBEGIN and %STPEND macros were added.

    Click the **Finish** button to register and run the stored process.

The stored process prompts you for the values and creates your report. The results of this stored process are shown in Section 3.2, "Running SAS Stored Processes."

3.5 Enhancing Your Stored Process Output

Once you have created a stored process, you might notice that the output uses SAS Default as the style. Unless you are using a client application, the output is always to a Web browser. You can control the appearance and the output location using the ODS options. ODS options are passed to the stored process prior to execution, based on the output.

ODS options control how results are handled, where results are sent, and even the image type and size. There are numerous ODS options available to assist with controlling the ODS output, but to control style and destination there are three to focus on: _ODSDEST, _ODSSTYLE, and _ODSOPTIONS.

There are shared prompts available in default installations to use immediately for ODSDEST and ODSSTYLE within the Samples folder.

3.5.1 Using Different SAS Styles

When you want to use a specific SAS style, you can change the _ODSSTYLE option. SAS provides a shared prompt to assist with changing the style.

In the following example, you are going to set a default style for the stored process you just created in the previous section.

1. Right-click the stored process you want to modify and select **Modify *<stored process name>***, where *<stored process name>* is the name of your process.

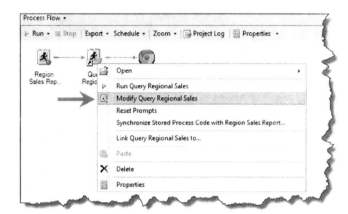

2. Go to the Prompts pane. Select **Sharing > Add Shared**.

3. Navigate to the default directory. There are a number of shared prompts available. Select **ODS Styles – Static**.

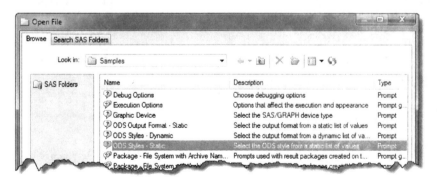

4. The ODS Style is added to the **Input Prompts** list. However, because the prompt is shared, you cannot make modifications to it. Select the prompt and then select **Sharing > Unshare**. A notice window appears with a warning about the prompt; you can ignore it.

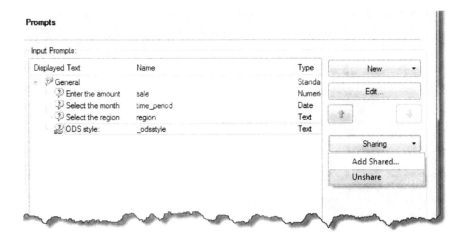

5. In the Edit Prompt window, change the prompt name to **Style** and select the **Hide from user** check box. This hides the prompt from view. If you want to allow the user to change the style, then do not select this check box.

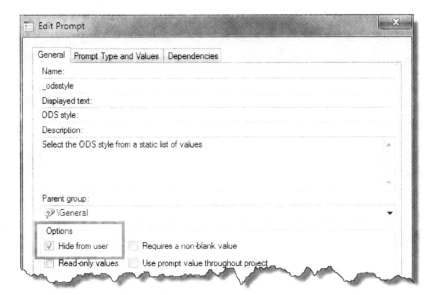

6. On the **Prompt Type and Values** tab, select a default style from the list. When you have finished adding the prompt, run a test to see the results. In this example, the font, text color, and background color are different.

3.5.2 Using a Cascading Style Sheet

Your organization might have a cascading style sheet (CSS) to ensure a consistent look to all Web sites. You can use that style sheet with your stored processes that output to HTML. To set the style sheet, create a prompt with the name _odsstylesheet and set the default value to equal something similar to:

```
(URL" "http://myserver/my_css_file.css");
```

The URL must be the Web location where your style sheet is stored. You need the complete path and complete filename.

 In some cases, a SAS BI client, such as SAS Web Report Studio, might not adopt any style overrides and defaults to the user-defined style.

3.5.3 Publishing Results to Multiple Devices

SAS Business Intelligence supports multiple output destinations, such as PDF, HTML, Microsoft Excel, and RTF. Keep in mind that the client application must support the destination you define. For instance, if you specify PDF as the destination, the stored process might produce unexpected results when run from the SAS Add-In for Microsoft Office or even SAS Web Report Studio.

To set the device, use the following ODS option:

```
%let _odsdest=device-type;
```

Some common device types are: CSV, EXCEL, HTML, PDF, RTF, and XML. Refer to the SAS customer documentation for a complete list of supported devices for your SAS version.

You can use the same method to create a default device prompt as was used in Section 3.5.1, "Using Different SAS Styles."

3.6 Working with SAS Stored Processes

So far, the stored process examples have been simple to help you understand the fundamentals. Stored processes can do more than display the output. In this topic, you will learn more about using macro variables and how to add more pizazz to your stored processes.

3.6.1 Understanding Macro Language Fundamentals

This topic introduces fundamentals of the SAS macro language as it applies to building stored processes. The macro language allows code to be re-useable, providing an increased amount of flexibility to programs. For complete information about the subject, see support.sas.com for SAS Press titles, SAS training, and other references.

3.6.1.1 Working with Macro Variables

A macro variable allows you to substitute a value in a SAS program when the program is executed. Stored process input parameters are supplied by the macro variables. To create a macro variable, you can use a prompt, %LET statement, PROC SQL, or CALL SYMPUT statement. In a SAS program, a macro variable has an ampersand (&) symbol in front of it when used in code.

LET statements and the resulting macro variables were demonstrated in Program 3.3-1: SAS Code with %Let statements.

3.6.1.2 Working with Global Macro Variables

In a stored process, you declare all input parameters as global. When you declare a global variable, an empty variable is created. If you do not declare the global variable manually within the code, or if it is not created automatically through the SAS Stored Process wizard, your stored process can generate warning messages and might contain other errors.

3.6.1.3 Double Quote the Variables

With the exception of numeric values, always use double quotation marks to resolve the macro variables values for use by stored processes. The quotes allow the values to translate successfully into the values selected by the end user or assigned by default. If you use single quotes, the macro values are not resolved and the stored process will not work as expected.

The following code demonstrates how to use a macro variable as a title in your stored process output. From our existing code, you can use the macro variables in the title so the users can see the selected values. When the user selects the variable from the input prompts, the values can be used multiple times in the code.

To add the existing macro variables to the title, insert a new TITLE2 statement and set the text in double quotation marks. Add the macro variables with any text you would like. In the following code, you can see that the TITLE2 statement was added and the global macro variables are used. When the quotes are added to the &SALE macro variable, it resolves as a character value.

```
/*=== Start the Stored Process ==*/
%STPBEGIN;
title "Customer Sales by Region";
title2 "Minimum order value $&sale.";
```

In Figure 3.6-1, the macro variable translated into the TITLE2 statement, based on the value that the user chose during the stored process execution, which was 6500.

Figure 3.6-1 Title shows the macro variable

3.6.1.4 Understanding Macro Logic

When you combine macro variables with macro logic, the stored process can make decisions and take actions based on the decision. For instance, in the first stored process you created, the user was allowed to enter a minimum order value. What if the data does not contain any orders over 100,000, and the user enters 120,000? The stored process would run and not return anything. The user would wonder if there was an error or if the stored process was just extremely slow.

By using the %IF/%THEN/%ELSE macro statements, you can code stored processes to generate meaningful end messages to users based on the following scenario.

1. To use these macro logic statements, you must add the macro start (%MACRO) and end (%MEND) statements so SAS knows you want to use the macro logic. In Program 3.6-1, the macro start and end statements are placed around the code, and the macro is named MakeReport. The macro name is how it is referenced later.

2. The code generates a temporary data set using PROC SQL. When PROC SQL runs, it generates an automatic macro variable (&SQLOBS.) with the number of returned observations.

3. Use this automatic variable for your test condition in the %IF/%THEN logic.

   If the &SQLOBS equals zero then use PROC PRINT to display a message to the user. Else, if the count is not zero then the data is available for the report.

4. In the final step, the %STPBEGIN/%STPEND stored process macros are placed around the macro name, MakeReport. The titles are placed before the macro because both apply to the output.

```
/*======================================================================*/
*ProcessBody;
%global region time_period sale;
 libname mylib meta library="Candy";


%MACRO MakeReport; /*== START MARCO ==*/
/*=== Use PROC SQL to query the dataset and create a TEMP dataset */
        proc sql;
                create table TEMP as
                        select region, name, date, sale_amount
                from mylib.candy_sales_summary
                where date ge "&time_period."d
                        and sale_amount gt &sale.
                        and region in ("&REGION");
                quit;


  /*=== The number of rows in the TEMP dataset is assigned to &SQLOBS. */
        %IF &SQLOBS. le 0 %THEN %DO; /*Send error message to user */
                data msg; MESSAGE="Values not found. Try again.";run;
                proc print data=msg noobs; run;
        %END;
        %ELSE %DO; /*Rows were returned; output the report */
                proc report data=temp nowd;
                column region name Date sale_amount;
                define region/group;
                define Name/group;
                define date/group format=monyy.;
                run;
    %END;
%MEND makereport; /*=== END MACRO ===*/


/*=== Start the stored process output and call the macro ===*/
%stpbegin;
                title "Customer Sales by Region";
        title2 "Order over $&sale.";
                %MakeReport; /* Call the macro */
%stpend;
/*========================================================================*/
```

Program 3.6-1 Using IF/THEN/ELSE logic

The user was searching for orders over $120,999. However, there were no sales over that amount for the month so the stored process then alerts the user.

Figure 3.6-2 SAS Stored Process output when no rows are returned

3.6.1.5 Integrating Multiple Selection Prompts into a Stored Process

Some of your stored processes might allow the user to make multiple selections. In the following sample stored process, the user can select only one region. It is easy to imagine that the user might want to see the report with more than one region.

You can update the Region prompt to present multiple values to the user, as shown in the following figure. However, the stored process code has to support this modification as well.

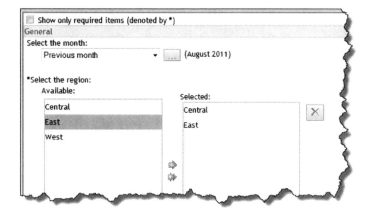

Figure 3.6-3 Allowing multiple selections for Region prompt

To understand how the change appears on the back end, you can review the stored process log. A log is produced with each stored process run. The log shows the values associated with each prompt, among other items. In the following program, the left log shows the results when the region prompt is set up for single values. There is only one REGION variable and it contains East. In the right log, the region prompt was set up to accept multiple values. Many more variables are created and passed to the stored process.

One Selection	Multiple Selections with One Item Selected	Multiple Selections with Multiple Items Selected
>>> SAS Macro Variables: REGION=East SALE=6500 TIME_PERIOD=01Jun2011	>>> SAS Macro Variables: REGION=East **REGION_COUNT=1** SALE=6500 TIME_PERIOD=01Jun2011	>>> SAS Macro Variables: REGION=East **REGION0=2** **REGION1=East** **REGION2=Central** **REGION_COUNT=2** SALE=6500 TIME_PERIOD=01Jun2011

Program 3.6-2 Partial log from three stored processes

The additional variables communicate not only the additional choices, but also the number of values. REGION_COUNT and REGION0 contain the number of user selections for this prompt. REGION1 and REGION2 contain the values the user chose.

You can accommodate these changes within the WHERE statement. Add some macro logic that allows the stored process to make a decision about what needs to be included based on the REGION_COUNT variable. When only one item is selected, as denoted in the center column above, the REGION_COUNT = 1 and there are no REGION1, REGION2 variables in existence. Therefore, code for this situation and use only the REGION variable. This code sends as many region values as it finds.

More information on automatic variables created through prompts is available in Chapter 4, "The Prompting Framework."

When adding these values with a macro, you need to use double ampersands so that the variable resolves correctly. Using the double ampersand technique, as in the following example with &&REGION&CTR, the macro is forced to resolve twice. First it resolves &&REGION to &REGION and &CTR to 1. The IN operator takes the combined &REGION1 to resolve to East.

```
/* ========================================================== */
%macro MakeReport;
proc sql;
      create table TEMP as
      select region, name, salemonth, sale_amount
      from mylib.candy_sales_summary
      where sale_month = "&time_period."d
            and sale_amount gt &sale.
            and region in
            ( %if &REGION_COUNT. = 1 %then %do;
                    "&REGION"
      %end;
            %else %do CTR = 1 %to &REGION_COUNT.;
                    "&&REGION&CTR."
            %end;
            );
      quit;
/* ========================================================== */
```

Program 3.6-3 Working with multiple values

The following figure shows the results from the stored process. Both the Central and East regions are populated.

Figure 3.6-4 Results from multiple regions

3.7 Tips and Tricks

SAS Stored Processes offer a lot of flexibility to report builders. For instance, a stored process can be used in other SAS BI clients, chained together to provide drill-down capabilities, modified to use HTML code, and to perform some other techniques that will be discussed in this section.

3.7.1 Chaining SAS Stored Processes

When developing reports, often data from a report leads users to want more details. The power of a stored process is easily having the information available. In the following figure, the summary report lists the top 10 orders for the year. The user clicks the order number to reveal the order details.

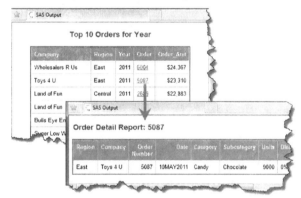

Figure 3.7-1 Chaining stored processes example

To create the link between the two stored processes, create a hyperlink in the parent stored processes and pass the variable information through the URL address to child stored process. You can get the URL address for the stored process by running it once in a Web browser.

 Navigate from the starting point http://server name: portnumber/SASStoredProcess/ do?_action=tree to quickly get the full URL path from the Internet browser tool bar.

The following figure shows an example from the Order Detail Report stored process. Once you have the address and know the prompt names, you can create the hyperlink using HTML code. In Program 3.7-1, the hyperlink is added to a temporary data set that is created as the stored process is made.

Figure 3.7-2 Capture the URL for a stored process

To create the hyperlink in the stored process, you can take advantage of the reserved macro variables (%_SRVNAME , %SRVPORT, and %_METAFOLDER) to recreate the stored process path. In Figure 3.7-2, the browser used %2F in place of the backslash (/) symbol. The URL address can contain only characters, so the PUT function is used to change the number to a character value.

The ORDER_URL is given a large length and format because the paths can become very long. This ensures that the hyperlink is not truncated, which causes the link to not work.

```
/* ======================================================================*/
%LET SP_NAME=Order_Detail_Report;
 libname mylib meta library="Candy";
proc sql outobs=10;
create table top10ords as
select saleyear
 , company
 , region
 , '"<a href=http://
&_srvname.:&_srvport./SASStoredProcess/do?_program="
||tranwrd(strip("&_METAFOLDER"), ' ',)||'+')||"&SP_NAME."
                        ||'&orderid='||strip(put(orderid,8.))
                        ||" target=_blank>"
                        ||strip(put(orderid,8.))||"</a>"
                              as order_url
           length=500 format=$500. label='Order Number'
      , sale_amount format=dollar12. label='Order Amount'
 from mylib.sales_candy_history
      where saleyear=&year.
 order by sale_amount desc;
quit;

Title "Summary - Top 10 Orders for &Year.";
proc print data=top10ords noobs label;
run;
/* ======================================================================*/
```

Program 3.7-1 Chaining stored processes code example

You might want to consider creating several stored processes that contain detailed information and adding the variable with the hyperlink to your data sets. Then, as you create other stored processes, the variable with the stored process link is already set up and ready to use. This is a time saver and makes creating reports with the chaining a snap.

3.7.2 Using HTML Code in SAS Stored Processes

Using HTML code for use in SAS Stored Processes provides developers with the ability to create an endless variety of forms and results, as well as the ability to create interactive content driven by SAS data. Included in this section is an example of printing a retrieved sales record from the sales transactions in a friendly HTML format. When using standard SAS code to print a single order, the layout is horizontal by default.

Figure 3.7-3 Default layout for SAS Stored Process

Using different SAS procedures such as PROC REPORT and using the ODS options can provide some layout options. However, Program 3.7-2 is an example of using the HTML code within a SAS Stored Process. Note that when using file _webout, the %STPBEGIN and %STPEND macros must be turned off and code for ODS HTML must be manually included. The ODS HTML statements must also include no_top_matter and no_bottom_matter ODS options.

```
/*=== Start the Stored Process ============*/
/*=== Define the library to the stored process ==========*/
 libname mylib meta library="Candy";

/*=== Use ODS HTML Statements instead of %STPBEGIN and %STPEND to use html code within the
_webout*/
/*=== no_bottom_matter = suppresses writing lower half of standard html code*/


 ods html body=_webout(no_bottom_matter
             title='Order Detail Report')
         path=&_tmpcat (url=&_replay)
         style=sasweb;

/*=== close the ods html output location to allow for writing to the _webout*/
 ods html close;

/*=== setting the order_number prompt value to global*/
```

```
 %global order_number;

/*=== use an empty data step to write out to the file _webout*/
 data _null_;
/*=== define the file _webout*/
 file _webout;

/*=== set the source dataset*/
 set mylib.candy_sales_summary;

/*=== filter the data table on the prompt value 'order_number'*/
 where orderid = &order_number;

/*===   create a custom measure to write out total sales $ with a format to dollar9.*/

        format total dollar9.;
        total=retail_price*units;

/*=== use HTML code within put ''; or put ""; statements*/
 PUT "<p>";
 PUT "<a href=http://www.sasbibooks.com>More Information</a>";
 PUT "</p>";
 PUT "<!--GENERAL AREA ============================================ -->";
 PUT "<table border=0 cellpadding=2 cellspacing=2 width=95%>";
 PUT "<tr valign=middle bgcolor=#B0B0B0 >";
 PUT "<td align=left colspan=5>";
 PUT "<a id=1><font size=2><b>Order Number: " orderid"</b></font></a>";
 PUT "</td>";
 PUT "</tr>";
 PUT "<!-- ============================= -->";
 PUT "<tr valign=middle bgcolor=#D3D3D3>";
 PUT "<td><font size=2><b>Name</b></font></td>";
 PUT "<td><font size=2><b>Region</b></font></td>";
 PUT "<td><font size=2><b>Category</b></font></td>";
 PUT "<td><font size=2><b>Product</b></font></td>";
 PUT "<td><font size=2><b>Units</b></font></td>";
 PUT "</tr>";
/*==There are three other rows created using the same code and format as above*/
size=2>" total"</font></td>";
 PUT "</tr>";
 PUT "</table>";
run;

/*=== no_top_matter = used at the bottom of the process to suppress the writing of the top
half of standard html*/

ods html body=_webout(no_top_matter
 title='Order Detail Report')
        path=&_tmpcat (url=&_replay) style=sasweb;
ods html close;

/*=== END OF CODE ===*/
```

Program 3.7-2 Using HTML code within a stored process

This is sample output from the stored process.

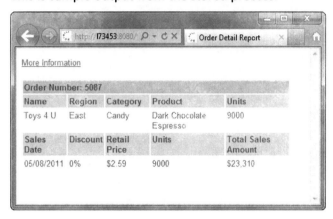

Figure 3.7-4 Using HTML Code in a stored process

When setting up the stored process, the %STPBEGIN and %STPEND statements must not be appended automatically by SAS Enterprise Guide. Always select the **No** button when you receive the following warning message.

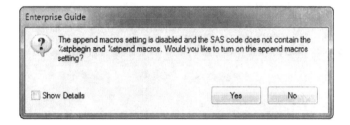

Figure 3.7-5 Warning message about %STPBEGIN/% STPEND macros

 Warning message for the appended macro codeThe %STPBEGIN and %STPEND macros option will always be turned on when modifying stored processes directly from SAS Enterprise Guide. It is important to double-check this option in the SAS Enterprise Guide wizard to ensure that it matches your stored process requirements.

Other examples for HTML code include using custom text and images within the graphical results, adding JavaScript functions, or implementing custom input forms.

3.7.3 Running a Stored Process in the Background

Some stored processes require more time to complete. However, the default behavior is for the processes to run in the foreground, which locks the open browser session from completing anything else. Moving the process to the background, the stored process disconnects from the browser session and allows the user to continue using the browser for other Web-based activities.

3.7.3.1 Modifying the Stored Process To Run in the Background

There are multiple mechanisms to run the process in the background. The simplest method is to modify the URL so that the &_action= parameter is set to BACKGROUND. For example, the following URL was &_action=update and after the change was &_action=BACKGROUND. You could click **Enter** to rerun the stored process.

Figure 3.7-6 Changing the &_action to BACKROUND

Another method is to create a hidden prompt for &_action that forces the stored process to always run in the background. This method requires that the stored process have at least one visible prompt and use only the package output.

After creating at least one other prompt, add a new prompt called _action. and hide the prompt from the user. Then set the default value to BACKGROUND as in the following sample.

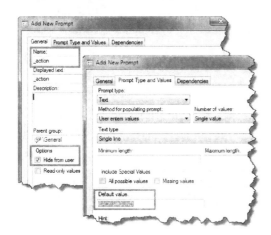

Figure 3.7-7 Creating an _action prompt

3.7.3.2 Checking On the Stored Process Status

After submitting the stored process to run in the background, the Web browser provides a status message.

Figure 3.7-8 Message indicates background job submitted

Using the SAS Information Delivery Portal, you can check the stored process status through an Alert portlet. You can also access the stored process package results from the portal. Refer to Chapter 10, "SAS Information Delivery Portal," for specific instructions on how to add portlets.

The Stored Process Alerts portlet displays all of the completed stored processes that ran in the background and that created package results. If you click on the stored process name, the result appears. Other actions are available, such a adding a bookmark, publishing, or e-mailing the result.

 Alerts appear when the stored process has finished executing and was set to generate a transient or permanent package.

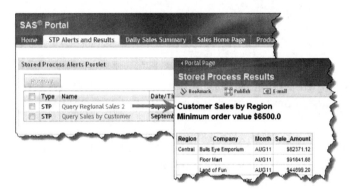

Figure 3.7-9 SAS Stored Process Alert portlet and results

3.7.3.3 Viewing the Stored Process Results

Packages generated from stored processes are typically stored in the SAS Content Server or WebDAV locations. In the SAS Information Delivery Portal, use the SAS Navigator portlet or the WebDAV Navigator portlet to view the results.

In the following example of the Personal Repository Navigator portlet, there are several packages available. These results are available in this path: **WebDAV > Users > <user name> >PR >MyDocuments**.

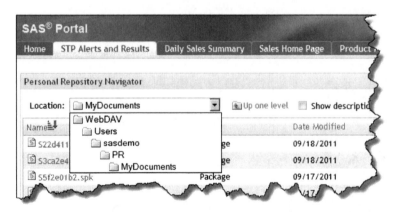

Figure 3.7-10 Storage location of stored process package results

3.7.4 Custom Output in SAS Web Report Studio

In SAS Web Report Studio, you cannot resize objects, such as graphs, that are generated from a stored process. Thus the stored process might require several printed pages, even if the output is displayed on one screen. When using SAS/GRAPH code, you can use GOPTIONS to control the size of the graphs.

In the following code sample, the GOPTIONS VSIZE and HSIZE have been added. With these options, you can control the graph size and how it prints on the page. Assuming that the margins are set to 1-inch on all four sides, the following program sets the values for VSIZE and HSIZE to 3 inches.

```
/* ===================================================================*/
*ProcessBody;
%stpbegin;

/* Add the GOPTIONS VSIZE and HSIZE */
goptions vsize=3 in hsize= 3 in;

SYMBOL1 INTERPOL=JOIN  HEIGHT=8pt      VALUE=DOT       LINE=1 WIDTH=1;
Axis1   STYLE=1         WIDTH=1         MAJOR=NONE      MINOR=NONE;
Axis2   STYLE=1 WIDTH=1 MAJOR=NONE      MINOR=NONE;

TITLE "Average Sales by Month";
proc gplot data=plotme;
 plot sale_amount*sale_month=region /
 vaxis=axis1 haxis=axis2;
run;
quit;
%stpend;
/* ===================================================================*/
```

Program 3.7-3 Controlling the chart size

You can adjust the margins. The VSIZE and HSIZE values depend on how many graphs you want to print on one page. You might need to try several different settings to achieve the best results.

3.7.5 Creating Custom Indicators in SAS BI Dashboard

The following options must be included for the graphical results to properly display within the SAS BI Dashboard.

- Include the following GOPTIONS within the stored process code.

  ```
  goptions gsfname=_webout gsfmode=replace device=png;
  ```

- Ensure that the %STPBEGIN and %STPEND macros are not included (automatically by SAS Enterprise Guide or manually) in the code.

- Set the Result Type to streaming.

- Get the full URL path from the Stored Process Web Application for the executed stored process.

- Use the Stored Process Web Application Tree view (http://server name: port number/SASStoredProcess/do?_action=tree) to quickly generate a full URL path.

- Within SAS BI Dashboard, create a Custom Graph Indicator. Copy the path generated above into the Image URL field.

3.7.6 Working with Metadata Libraries

The preferred method to define libraries for use by stored process code is to create META librefs. Using code similar to the following, modify the *libref* value with an appropriate name and replace *value* within the @name area to match the exact library name defined in your server list. Refer to Chapter 3, "SAS Enterprise Guide," for more information about metadata libraries.

```
libname libref meta library="value";
```

For example, to indicate the Candy metadata library, add the following to your stored process code.

```
libname mylib meta library="Candy";
```

3.8 SAS Administrator Tasks

The SAS administrator ensures that data and libraries are available for the stored process and sets up the source code repository. The following sections explain how this is accomplished.

3.8.1 Accessing Pre-Assigned Metadata Libraries

There are several strategies for making data available to stored processes. The preferred method is for developers to include META LIBREFs as described in Section 3.7.6, "Working with Metadata Libraries."

Another option available is to pre-assigning libraries. When libraries are pre-assigned, the SAS server environment initializes the connection using the identity that runs the spawner service. In typical installations, the SASSRV account performs this action; therefore, the SASSRV account must have READMETADATA access authorization on the libraries in SAS Management Console.

This can be accomplished by two methods:

- Add the LIBREF code to the AUTOEXEC file. Saved as autoexec.sas within the root SAS program folder, this program is automatically run when SAS is started.

- In SAS Management Console, modify the library options and then update the server startup script or the configuration file using SET commands.

 Do not pre-assign a library that requires user information to authenticate, such as with a RDBMS data source.

At a minimum, users must be granted READMETADATA and READ access to the library and associated data tables. In the following figure, access for the Candy organization's DATA folder is granted to the user group Category Group for Candy. This access is then inherited by all libraries stored in the DATA folder.

Figure 3.8-1 Set permissions in SAS Management Console

3.8.2 Setting Up a Source Code Repository

SAS Stored Processes exist as SAS programs (.SAS) in a physical folder location that must be accessible by the SAS server environment. For stored process developers to write and modify these programs, a storage location must be defined and maintained. Developers must have Write/Read/Modify access to the physical folder path.

 If multiple organizations have groups generating stored processes, create multiple source code locations.

Once the folder is created on the server, the location must be defined within the metadata for the developers using SAS Enterprise Guide. Use the following steps to add a location.

1. From SAS Management Console, select the icon for **New Stored Process.**

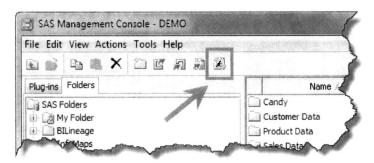

2.   In the New Stored Process window, do the following:

   a.   Because you are not going to create a new stored process, type a single letter in the **Name** field to move past the window.

   b.   In the Execution window, select the appropriate server and click the **Manage** button.

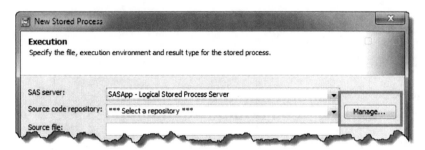

   c.   Use the **Add** button to add all of the necessary code repositories. Click **OK** to continue.

   Any libraries that you have added are now available from the Source code repository drop-down list, as shown in the following figure. For this example, the C:\SAS\data\Candy folder was defined.

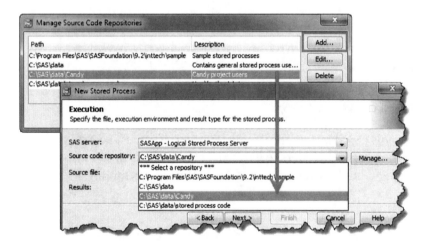

3.   Click the **Cancel** button to exit the New Stored Process wizard.

Chapter 4

The Prompting Framework

Chapter 4
The Prompting Framework

Prompts are commonly seen on the Web, allowing users to enter information, search through data, or interact with a tool. Consumers can apply for loans, search for automobiles, or choose preferences using forms made from multiple prompts.

In SAS software, prompts provide developers with the capability to produce more dynamic content with less effort and fewer resources. Developers can leverage prompting functionality by enabling one program to meet multiple needs rather than using smaller programs that must be modified or duplicated for various uses. Reusing material improves productivity and efficiency by offering consistent and organized items across multiple products.

Users also have a friendly interface that offers data validation and removes the need to know how to modify SAS code. Prompts guide users through a report, automatically generating code with values they have selected. Reports can use the prompt value to do things such as filter the data, specify titles, or stylize the report.

As with many components, prompts are the most valuable when the interface is user friendly and provide the results users need. This requires an understanding of not only creating prompts, but how to store and share them with other components of SAS Business Intelligence system.

4.1 Getting Started

The following sections describe the underlying technology of prompting used across multiple BI clients, including SAS Stored Processes and SAS Enterprise Guide.

You can create prompts from several locations but you can only share them using SAS Management Console. When sharing prompts, multiple applications are able to use the same prompt.

The following table provides an overview of where you can create a prompts and where you can share a prompt.

Tools	Create Prompts	Create Dynamic Prompts	Create Dependent Prompts	Share Prompts
SAS Management Console	Yes	Yes	Yes	Yes
SAS Information Map Studio	Yes	Yes (from data added to the map)	Yes	No

Tools	Create Prompts	Create Dynamic Prompts	Create Dependent Prompts	Share Prompts
SAS Enterprise Guide	Yes	Yes (from data tables only)	No	No
SAS Web Report Studio	Yes	Yes (only from data selected for the report section)	Yes (only from data selected for the report section)	No
SAS BI Dashboard	Yes	No	No	No

Table 4.1-1 Tools that support prompt creation

Prompts created in SAS Enterprise Guide are available only within that specific project. The only caveat is when the project flow or specific task using a prompt is converted into a SAS Stored Process. When this prompt is converted, the prompt moves to the Stored Process system and is no longer associated with the prompt in SAS Enterprise Guide.

If users create a prompt in SAS Web Report Studio or SAS Add-In for Microsoft Office the prompt is only available when the filter is run for the specific report.

Indicator interactions can be developed in SAS BI Dashboard to simulate dependent and dynamic prompts. Refer to Chapter 9, "SAS BI Dashboard," for more information how to implement interactions.

4.1.1 Quick Tour

The interface to create new prompts (used in this section) is identical in SAS Enterprise Guide, SAS Stored Processes, SAS Management Console, and SAS Information Map Studio.

- From SAS Management Console, select **Properties** on an existing stored process. Then, from the **Parameter** tab, select **New Prompt**.

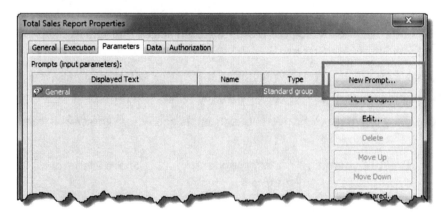

- In SAS Information Map Studio, select **Manage Prompts** from the **Tools** menu. Then select **New** on the Manage Prompts window.

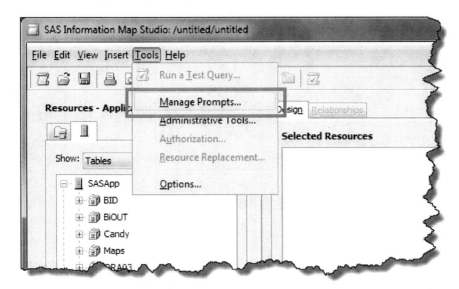

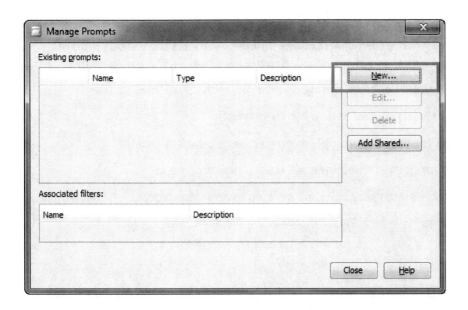

- From SAS Enterprise Guide, select the **Add** button within the Prompt Manager window.

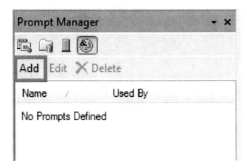

- From the SAS Enterprise Guide Create New SAS Stored Process wizard, select **New Prompt** from the **New** drop-down box on the **Prompts** page.

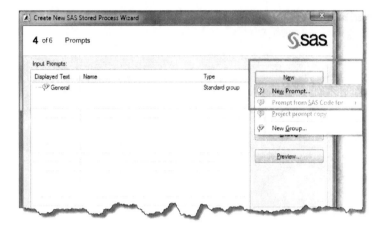

4.2 Understanding Prompts

Prompts can be used to meet a variety of your users' requirements as there are various prompting techniques, types, and functionalities. The following sections define how to use the various prompts, show where you can create the prompts, and give a detailed overview of each individual prompt.

You can further improve the user's experience by setting up prompts with capabilities such as:

- a dynamically derived list of values, referred to as *dynamic prompts*
- one prompt driving the values for the next prompt, referred to as *dependent prompts*
- groups of prompts organized for quicker user entry, also called *prompt groups*
- prompts that drive what other prompts are displayed, defined as *selection groups*
- creating prompts for reuse in multiple locations, otherwise known as *shared prompts*

4.2.1 Using Dynamic Prompts

Dynamic prompts point to data sets as the source values for the prompt. This is useful for changing values, such as the list of customers that exists in a weekly data table.

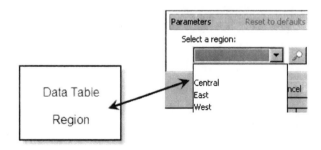

Figure 4.2-1 Dynamic prompts

In Figure 4.2-1, the list of choices for the region prompt are dynamically generated from the data table each time the prompt is displayed. The distinct list of regions is retrieved each time the prompt is displayed to a user.

When to use	Static prompts contain a list of values that is managed by an individual. If this list of values requires constant modifications and updates, using a dynamic prompt is the better option. For example, a static prompt with a list of customers would need modification any time new customers are added. Reduce the management of this prompt by pointing to a data source that is either the raw table or a refreshed summary table with the list of valid customers.
When not to use	Each time the prompt screen appears, it runs a query to select the distinct values within the source table. Because of this, users might experience a delay before they are allowed to make a selection. Therefore, you should not run this for prompts that do not require a dynamic list. For instance, for severity, the possible values (Critical, Major, Minor, Information) might never change.

When using dynamic prompts, if the data table source is extremely large, the response time to view the prompt might be long. If this is the case, you can create a summary table containing the list of distinct values needed for the prompt. Update this table as part of the overall extract transform load (ETL) process to ensure that it contains current information. Implementing an index on larger summary tables further improves its response time.

4.2.2 Using Dependencies between Prompts

Also referred to as *cascading prompts*, *dependencies* between prompts means that the first prompt selection is used to subset the values displayed in the second prompt. This is extremely useful if the second prompt list of values is extremely large, which causes the users first to see only the top 20 values and opt to view additional values.

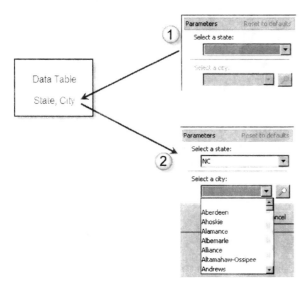

Figure 4.2-2 Cascading prompts

In this example, the data table provides two columns: State and City. The first prompt is a static prompt listing all US states. When users select a state ❶, the value is used to generate a list of cities for that state and returns those values to the second prompt ❷.

When to use   A common use for cascading prompts is on geographic data filtering, for instance from state to city. The list of cities would be extraordinarily large if it included all cities in the data set. Using state to filter city and then display only those cities within the selected state keeps the interface user friendly.

When not to use   If the second prompt is not a subset of the first, then using the dependency capability does not provide any value.

4.2.3 Using Prompt Groups

Prompt groups are prompts that are commonly used together, creating a usable interface for users. In this example, three prompt groups exist: **Location**, **Automobile,** and **Output Options**, each with a set of related prompts.

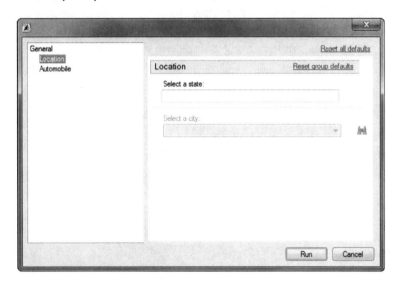

Figure 4.2-3 Grouped prompts

When to use   Prompt groups are helpful when organizing and sharing complex dependent prompts or when the data must use all prompts to be successfully queried. In the State – City example, grouping these prompts makes sense. Querying on City will cause incorrect results because the same city name exists in multiple states.

When not to use   SAS Web Report Studio cannot directly use grouped prompts. A workaround is to call a stored process that uses the prompt group.

                         You cannot group project prompts from SAS Enterprise Guide.

4.2.4 Using Selection Groups

Selection groups allow the user to select which prompt group to choose.

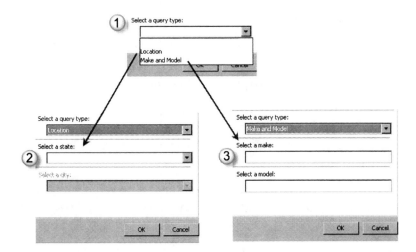

Figure 4.2-4 Selection group prompts

In this example, the user first is prompted to decide how they want to interact with the data. If the user selects the **Location** option in the **Select a query type** ❶ prompt, the State and City prompts appear ❸. Selecting **Make and Model** changes the prompts displayed so the user can then enter a make and model. This flexibility allows a single report to address even more user requirements.

When to use	Users need to first select which option group to display, and then the prompt changes, based on their selection, with subsequent prompts and filtering. For example, the user searches for an automobile to purchase. She can choose to search by the city the car is in, by make and model, or by price. If she selects the make and model option, the dependent prompts for make and model appear for further selection.
When not to use	SAS Web Report Studio cannot directly use selection group prompts. The workaround is to call a stored process that uses the prompt group. SAS Enterprise Guide project prompts do not provide the interface to group prompts together.

 Groups of prompts can be used only in SAS Information Map Studio and SAS Web Report Studio through an attached stored process.

In stored processes, creating the flexible code for various query scenarios can become complex. Also, selection groups cannot be shared between stored processes or information maps. You must create selection groups for each location in which they are used.

4.2.5 Using Shared Prompts

Individual prompts can be shared across all BI clients by creating and sharing them within the SAS Management Console New SAS Stored Process wizard.

You can create prompts in several different locations, but they can be shared only using SAS Management Console. Shared prompts are then available for multiple applications to use the same prompt.

This mechanism allows you to centrally manage the prompt text, the required formats, and so on.

When to use   Sharing prompts improves productivity. If the State–City dependent prompt is used in SAS Web Report Studio, stored processes, and SAS Add-In for Microsoft office, having the definitions in a single place makes management of changes much simpler.

When not to use   If the prompt is used in only one location, then there is no value with stepping outside of the tool it is used in to create it within SAS Management Console.

You can modify the prompt name, description, and displayed text without affecting the shared prompt definition in the metadata.

4.3 Creating a New Prompt

1. In the Add New Prompt window, do the following to create dynamic prompt. To create the Customer prompt (referred to in Chapter 6, "SAS Information Map Studio") type `CustomerPrompt` in the **Name** field and **Select one or more customers** in the **Displayed text** field.

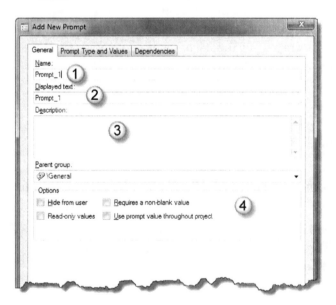

	Field Name	Description
1	**Name**	Must not contain any special characters or blanks.
2	**Displayed text**	The information the user sees, examples are: **Select the State** or **Choose one or more customers**
3	**Description**	This appears to users as subtext. It is useful when more information is needed to assist the user with interacting with the report. Example: "Select the type of query you would like to run. Location will provide prompts for State and City, while Make and Model will prompt you for that information."
4	**Options**	These are additional prompt options used to customize how the prompt interacts with the user and with the report. **Hide from user**: The prompt does not display to the user. This initially seems to contradict the reason you create prompts in the first place; however, this is commonly used by developers or administrators to create macro values applicable to a single report behind the scenes. An example is to set the reserved macro _odsstyle for a stored process. This macro sets the color schemes for the output. The user would not see this prompt or be able to modify it, but the report developer could modify it in the prompt GUI without changing the underlying SAS code. **Read-only values**: Similar to the **Hide from user** option, this sets a prompt to a prespecified value. Users cannot modify the value, but they are allowed to see it. For example, an organization specifies that users cannot query data older than one year, due to the size and performance requirements of the resulting query. A prompt with this option selected would allow users to see that this filter is present; however, they cannot interact or modify the filter itself. **Requires a non-blank value**: The user cannot go forward with running the report until the prompt has a value selected. Consider stored processes; an example of when to use this option is for cases when the code will return an error message and not run successfully if the user runs the report without selecting a value. **Use prompt value throughout project**: This option appears only for prompts within SAS Enterprise Guide. You can choose to use this prompt value for multiple tasks and for code within the project by selecting this option. Otherwise, the prompt is used only for specified task and code elements.

2. After entering values on this tab, move to the **Prompt Type and Values** tab. Continuing with the information map example, choose the populating prompt method **User selects values from a dynamic list** and select **Customer** for **the Formatted (Displayed) Values** data item.

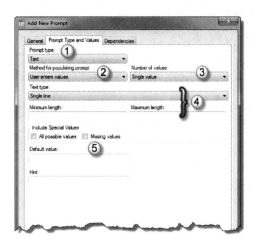

1	The **Prompt type** drop-down option is where you select the type of prompt. The options are listed and detailed in Section 4.7, "Quick Reference for Prompts."
2	**Method for populating prompt** specifies how the prompt values will be generated. • If users can enter their own values, then the prompt allows the user to type in the value and to see specific options based on the prompt type. For example, with the prompt type date, there is still a drop-down list with options such as **Tomorrow** or **N days ago,** as well as a calendar widget. But the prompt also would allow the user to manually **type** in "July 20, 2011". • Static lists are pregenerated values that the prompt displays to the user. When you create the prompt, you can manually type in available values or you can generate the list of values off an existing data table. The list does not update, so if an additional value is needed, you must modify manually through the prompt graphical user interface (GUI). An example of this option is detailed in Section 4.3.1, "Populating the List of Available Values Using a Data Table." • For prompts that require a changing list of values, you will find the dynamic list option valuable because of the reduced administration requirements. In Section 4.3.2, "Populating the List of Available Values Dynamically," you can learn where and how to use the dynamic list option.
3	If the user can select one or more values within a prompt, the **Number of values** option is modified accordingly.
4	In the preceding display, the prompt type **Text** is selected. Therefore, the available options include text type, length, and special character handling

	options. This entire area changes based on which prompt type is selected.
5	The default value entered is automatically assigned to the prompt selection. If the users can interact with the prompt, meaning it is not hidden from view or read-only, they can modify this value.

4.3.1 Populating the List of Available Values Using a Data Table

When you select the static list method of generating values for the prompt, you can manually type values for use in a prompt. However, it is much more efficient to upload the distinct values from an existing data table. When adding values in this manner, you remove the possibility of missing a value that exists in the table or inserting a value with a typo, which will result in zero records returned.

An example of using this is option is when generating a static list of US states. Administration of this list is minimal, because the number and names of US states very rarely change. Typing in the values of 50 states is time consuming, so getting the values from an existing data table reduces development time.

1. On the **Prompt Type and Values** window, select **User selects values from a static list**.

2. Click the **Get Values** button.

3. Based on which tool you are using, the Get Values window uses different methods.

    • From SAS Enterprise Guide and SAS Management Console, you can select a data table by clicking the **Browse** button.

    • From SAS Information Map Studio, you can use a data item that exists in the information map or you can select a separate data table.

4. After selecting the data source, the interface automatically populates the **Column** drop-down box with the list of columns that can be used by the prompt. Note that if, for example, you selected the prompt type **Date**, the list includes only date-formatted columns.

    For the prompt type **Text**, the Get Values window in the following the example shows the column selection **Two-letter abbrev. for state name**.

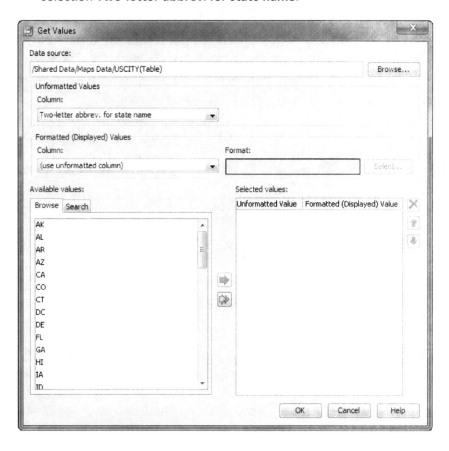

The **Formatted (Displayed) Values** drop-down box lets you select a different column which the user then sees. This is valuable when you have two different columns in your data source, such as STATENAME and STATECODE, and would like the user to see STATENAME, but you want the value passed to the query as STATECODE.

5. Move values from the **Available values** box to the **Selected values** box using the arrows.

6. After saving the prompt, you can see a drop-down box with the list of US states.

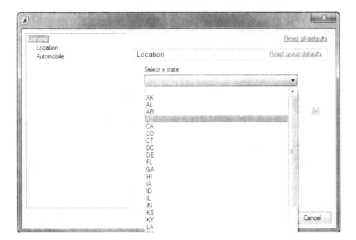

4.3.2 Populating the List of Available Values Dynamically

When creating a new prompt, select **User selects values from a dynamic list** from the **Method for populating prompt** list. Then click **Browse** to choose the data source.

• It is important to note that the data source must always be accessible to anyone you think will access this prompt.

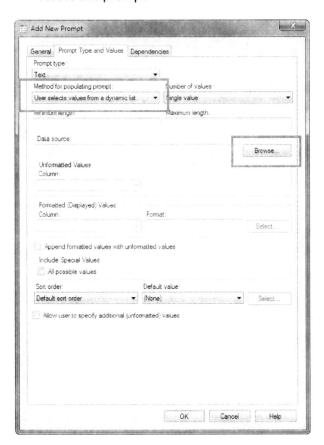

Figure 4.3-1 Select dynamic list to populate prompt

You can select an information map or a data set defined in the metadata.

You should choose a table that does not take long to return a SELECT DISTINCT value result; otherwise, the performance of the prompt can be impacted for all users.

4.4 Grouping Prompts

In Section 4.3, "Creating a New Prompt," you learned about several mechanisms available to group prompts together. This technique provides users with more intuitive and organized prompt screens.

New groups are created for stored processes, through SAS Management Console or through the SAS Enterprise Guide Create New SAS Stored Process wizard. To use groups from information maps, a stored process can be used to supply the prompting screens.

After groups and prompts are created, they are organized within the prompt structure. In the following example, there is also a transparent group called Single Prompts that helps to keep all the shared, non-dependent prompts in a single list.

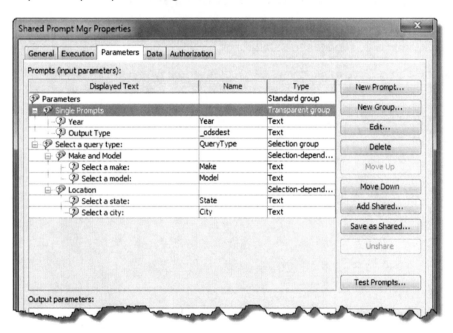

Figure 4.4-1 Grouping prompts

You will notice that the Single Prompts is a transparent group. This is useful to developers interested in keeping prompts organized; however, this setting does not make the prompts transparent to the users. Users will see all groups unless the option **Hide from user** is selected. For more information, refer to Section 4.4-2, "Creating a New Group Prompt." Notice that there are two selection group prompts that allow the user to select how to query the data either through **Make and Model** or through **Location**. Then there are dependencies between make and model, as well as between state and city.

4.4.1 Open the Add New Group Window

The interface to create new groups is identical in SAS Management Console and for SAS Stored Processes.

- From SAS Management Console, select **Properties** on an existing stored processs. Then from the **Parameter** tab, select **New Group**.

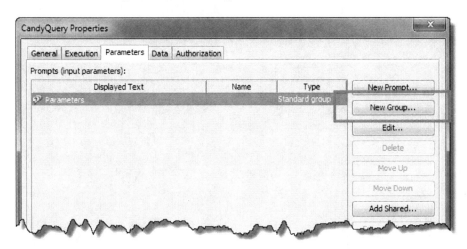

Figure 4.4-2 Add group prompt for SAS Management Console

- In the Create New SAS Stored Process wizard, select **New>New Group** from the Prompts page.

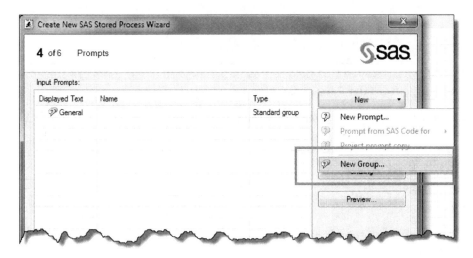

Figure 4.4-3 Add group prompt from SAS Enterprise Guide

4.4.2 Creating a New Group Prompt

The New Group window is identical for both the Create New SAS Stored Process wizard and SAS Management Console interfaces.

Figure 4.4-4 Creating a group prompt

1	Group types are explained in Section 4.4.3, "Group Types."
2	The information entered in the **Displayed text** field represents the menu name users can select or the name administrators see from the **Parameter** window.
3	Descriptions offer additional information on what the group does; however, this is not displayed for end users.
4	**Parent Group** allows you to specify where this new group resides.
5	The **Hide from user** option is available for all group types. When selected, end users are unable to view the entire group. **Requires a non-blank value** is available for selection groups. This requires the user to complete the selection prior to running the report.

4.4.3 Group Types

- Standard group

  Standard groups are used to organize prompts into more usable prompt screens. In the following example, the three standard groups are Location, Automobile, and Output Options. Each item selected on the left panel modifies the available prompts in the main window.

Figure 4.4-5 Standard group example

- Transparent groups

  In the following example, a transparent group for output options has been created so that administrators can see that these prompts are similar.

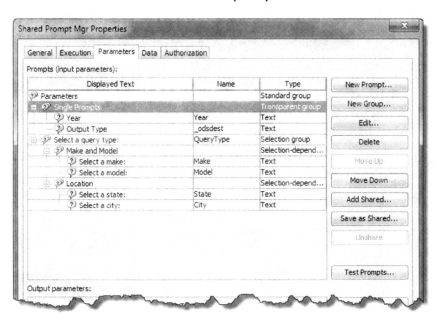

Figure 4.4-6 Transparent groups

However, when the user runs the prompts, the group is not seen.

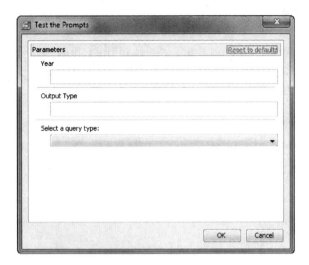

Figure 4.4-7 Transparent groups not seen by user

- Selection group

  Provides the user with the ability to choose the group of prompts to enter. In the following example the users can choose whether to only query data on **Location** or on **Make and Model**.

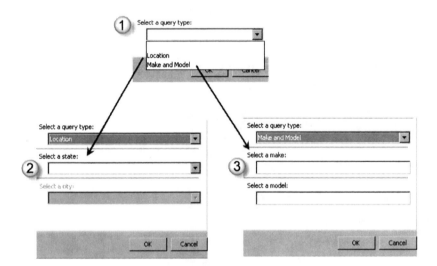

Figure 4.4-8 Selecting the next set of prompts to display using Selection Groups

4.4.4 Creating a Selection Group Prompt

As mentioned in the prior section, selection groups offer additional flexibility for users by allowing them to choose the next set of prompts before running the report. Creating a selection group prompt involves defined dependencies between the users selection and the available prompts. Included are the steps to create the example seen in Figure 4.4-8.

1. Choose the Group type Selection Group. Name the group Querytype and type Select a query type in the Displayed Text field.

2. On the **Selection-Dependent Groups** tab, select the **New Group** button.

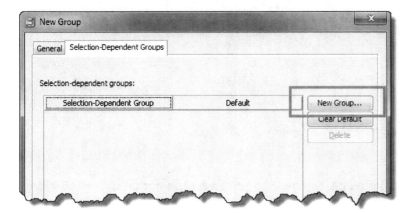

3. Define two groups, one for Location and one for Make and Model. In the following figure, the Location group properties are shown.

4. After entering these two selection groups, the New Group window appears as shown in the following figure.

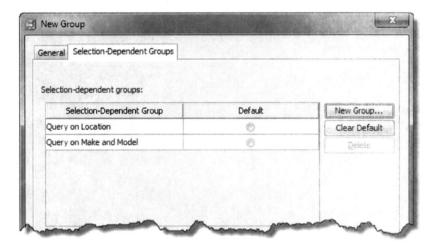

After completing the selection group definition, you need to edit each prompt that will exist within the selections.

The **Parameter** tab shows multiple groups but you must then reorganize the prompts for Location and Make and Model to show only when the user selects the new option to either **Query on Location** or **Query on Make and Model**.

5. You can quickly move these prompts by editing them and changing the parent group. In this example, select the **State** prompt and click **Edit**.

6. Change the parent group from **Parameters\Location** to **Parameters\Select a query type:\Location**.

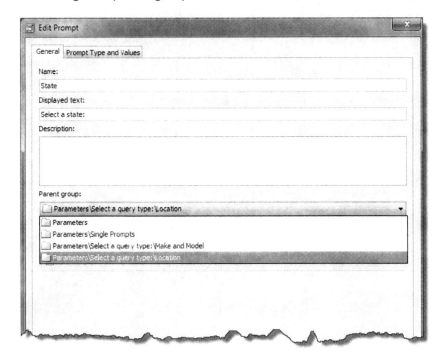

The **Parameter** tab looks like the following figure.

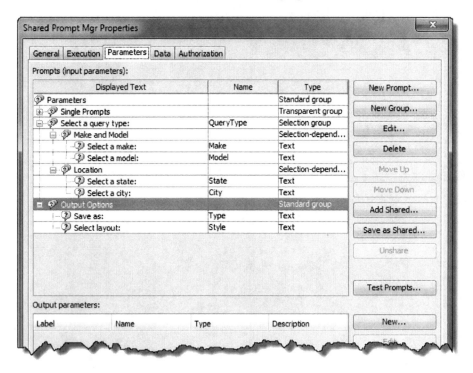

As shown in Figure 4.4-8, the user then can select the query type and be shown either prompt group.

4.5 Shared Prompting

There are a few tricks to working with shared prompts since they are ideally available to many users. The following topic explains how to setup and create shared prompts.

4.5.1 Set Up a Folder Structure for Shared Prompts

Before creating a shared prompt, you must determine the metadata folder structure where the shared prompts are saved. Shared prompts might be placed in a single folder or have different folders, based on project or organization security structures. You must grant read metadata authentication on this folder; otherwise, the users cannot access the shared prompts.

The simplest implementation is to create a new folder within **Shared Data**, which grants all SAS users read metadata access by default.

Figure 4.5-1 Creating a new folder

4.5.2 Set Up a Central Starting Point

You should create a single location to create and share prompts from SAS Management Console. This will also provide you with a quick way to access and edit previously created shared prompts.

1. From SAS Management Console, select the **My Folder** location within the SAS folder structure and open the New Stored Process wizard.

2. Type in a name, such as Shared Prompt Mgr.

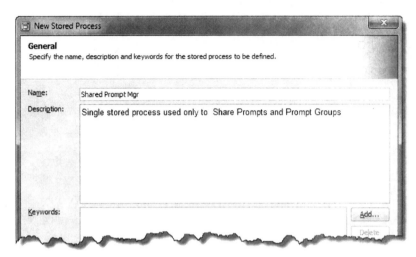

3. Specify the stored process server, the source code repository, and the source file name.

4. Save this stored process.

From this point forward, simply right-click on the stored process in the **My Folder** location and choose **Properties**. A parameter window appears where you can create and share prompts.

4.5.3 Creating a Shared Prompt

After creating the central starting point for shared prompts, open the Properties window.

1. Return to the folder where you started the stored process, right-click on the **Shared Prompt Mgr** stored process and select **Properties**.

2. Click **Parameters**. From here, you can add and edit prompts and prompt groups.

3. Click **New Prompt** and create a region prompt.

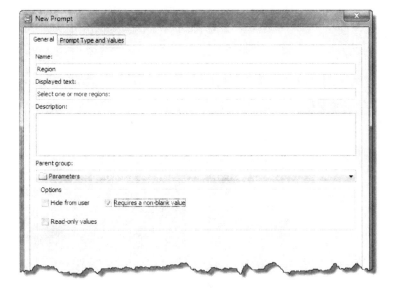

4. Move to the **Prompt Type and Values** tab. Because the organization has stated that the region values never change, select the static method for populating the prompt and allow users to select multiple values. As seen on the following screen, these values have been chosen. Then select the **Get Values** button.

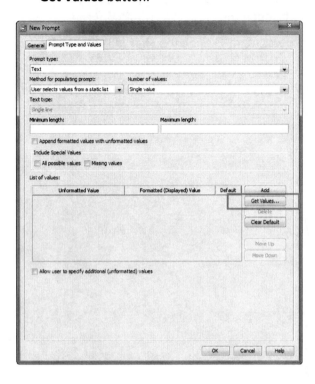

5. On the Get Values window, select the **CANDY_CUSTOMERS** table and choose the **Region** column. Then move the regions from the **Available values** list to the **Selected values** list, as seen in the following figure.

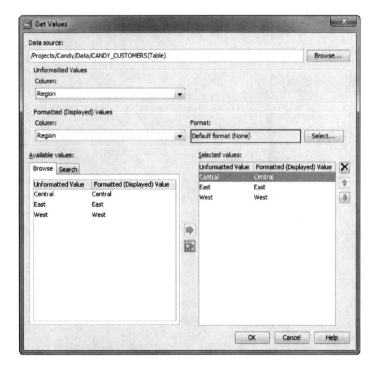

6. Choose **Central** as the default value by selecting the check box next to that row and save the prompt.

7. On the **Parameters** tab, verify that the new region prompt is highlighted and select **Save as Shared**.

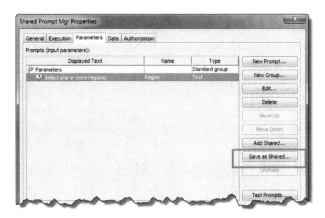

8. Navigate to the **Shared Prompt Folder** location and save the new prompt as Region. Now the prompt is ready to add to various information maps and stored processes.

4.6 Tips and Tricks

In addition to creating and sharing individual or grouped prompts, you might want additional functionality or be interested in working around some of the limitations.

4.6.1 Building Dependent Prompts

After you modify the second prompt in the selection group so that the user selects the values from a dynamic list, the **Dependencies** tab appears.

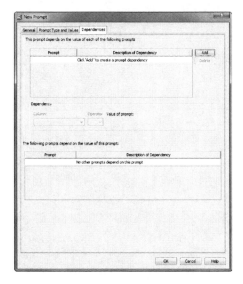

Figure 4.6-1 Dependencies tab

1. Select the **Add** button to create a filter on the value list based on the first prompt. In this example, choose the **Select a state** prompt to create the dependency.

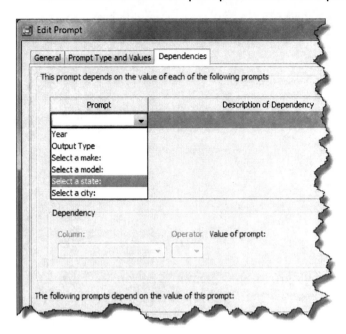

2. Modify the dependency conditions by changing the **Column** or **Operator** values, if needed.

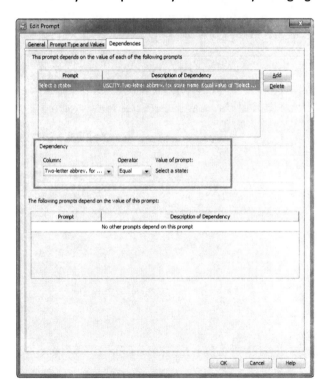

4.6.2 Creating Prompt Groups for SAS Web Report Studio

Information maps cannot create prompt groups; however, you can add a stored process that already has the grouping set up and include the stored process to the information map to display grouped prompts to SAS Web Report Studio users.

The following steps implement a grouped prompt in SAS Web Report Studio. The grouped prompt queries the user for the company name and a date range from a SAS Web Report.

1. Create a stored process, either from the Create New SAS Stored Process wizard in SAS Enterprise Guide or in SAS Management Console. Define the prompt groups as explained earlier in this chapter.

> The blank stored process code needs only one line to pass the parameters from the stored process prompt to the information map: `*PROCESSBODY;`

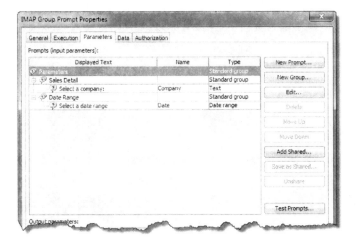

As we will mention in Chapter 6, "SAS Information Map Studio," the stored process must be registered to run in the SASApp – Workspace Server.

2. Create an information map, select the data table to query, and add the stored process. (Refer to Chapter 6, "SAS Information Map Studio," for detailed instructions.)

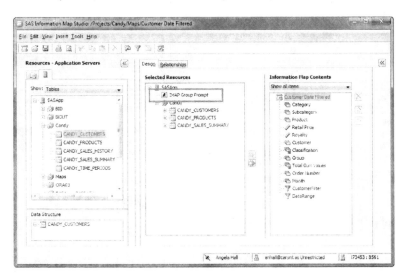

You might be unable to test the information map at this point; however, the report continues to run correctly in SAS Web Report Studio.

3.   Note that despite the fact that the stored process was added to the information map, the map is not filtering the data based on the user selections. You must create two filters for each prompt in this example and use the macro values generated automatically by the prompting framework.

4. Because these filters must always be included in the information map to return the filtered results, add these to the **General Prefilters** tab. This tab is located within the information map properties (found in the menu **Edit** -> **Properties** -> **Information Map**).

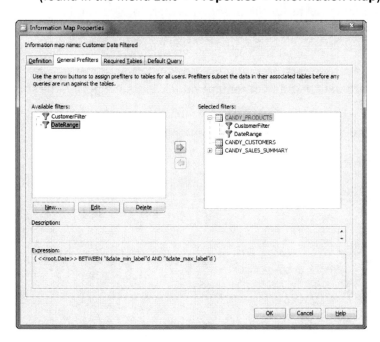

5. When you use this information map from SAS Web Report Studio, the user then sees the grouping of prompts.

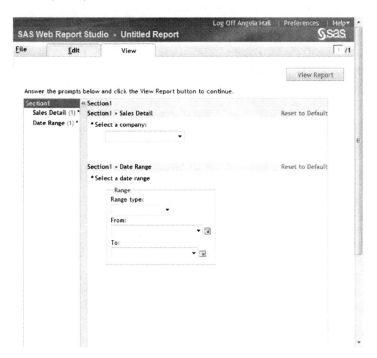

4.6.3 Making a Filter Optional

When using shared prompts from information maps, the prompt option **Requires a non-blank value** must be selected. When the method of generating values for the prompt is a static or dynamic list, an additional option in the **Include Special Values** area is available, called **All possible values**.

 This addresses the error message: The prompt *abc prompt name* does not require a value. SAS Information Map Studio cannot use prompts that do not require values. Please select a different prompt.

The following figure shows a before and after view of this special value option being selected.

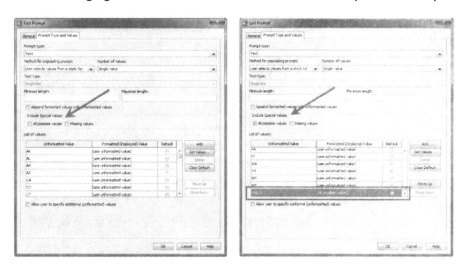

Figure 4.6-2 Optional filters

The user then sees the **Formatted (Displayed) Value (all possible values)** available in the prompt, as seen in the following figure.

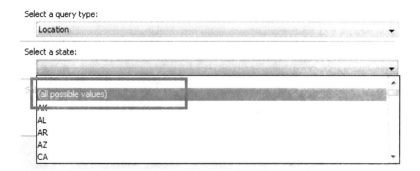

Figure 4.6-3 Example of all possible values

4.7 Quick Reference for Prompts

Each of the prompt types in the following table is available when you are defining a prompt. When a prompt is created, it generates a set of macro variables that can used by the report.

Prompt Type	Description	Available Macros
Text	Enter the value in a text box or select one or more text values. Text Prompt Example	Prompt_name If more than one option is enabled: Prompt_name Prompt_name_count Prompt_name0 (which equals &prompt_name_count) Prompt_name1 .. Prompt_name*n*
Text Range	Enter two values in side-to-side text boxes. Once entered, the prompt automatically tests either the alpha or numeric order to ensure that the lesser value is entered first. Text Range Prompt Example From: To:	Prompt_name_min Prompt_name_max
Hyperlink	Include a hyperlink. Also provides a box for link text. Note: This prompt is available for stored processes only through the SAS Enterprise Guide Stored Process wizard.	Prompt_name - which includes the link text Prompt_path - includes the URL

Prompt Type	Description	Available Macros
Numeric	Enter the value in a text box or select one or more numeric values. Once entered, the prompt automatically tests for integers. The prompt is similar to the Text prompt. However, if nonnumeric values are entered, an error message is generated, such as for the value "a" below. 	Prompt_name If more than one option is enabled: Prompt_name Prompt_name_count Prompt_name0 (which equals &prompt_name_count) Prompt_name1 .. Prompt_name$n$
Numeric Range	Enter two values in side-to-side text boxes. Once entered, the prompt automatically tests for integers and verifies that the range is valid. The prompt is similar to the Text Range prompt and provides the same validation checking as the error message shown for the numeric prompt above.	Prompt_name_min Prompt_name_max
Date	Enter the value by using a calendar prompt. Also provides the user the option to select a relative term (such as **yesterday** or **N days ago**). 	Prompt_name - returns results in date9. format. (01Apr2011) Prompt_name_label – returns full date (April 01, 2011) Prompt_name_end – if type of prompt selected is Week, Month, Quarter, or Year Prompt_name_rel – the relative term the user selected within the prompt

Prompt Type	Description	Available Macros
Date Range	Provides two date boxes with the same options as for the Date prompt. Once entered, the prompt automatically tests for valid date ranges. Also, range types are allowed for quick entry in the two date boxes. 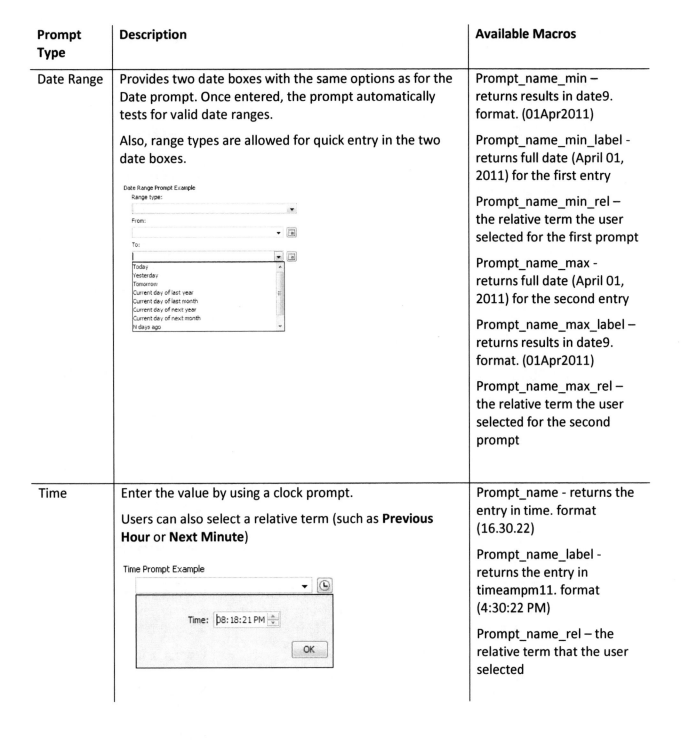	Prompt_name_min – returns results in date9. format. (01Apr2011) Prompt_name_min_label - returns full date (April 01, 2011) for the first entry Prompt_name_min_rel – the relative term the user selected for the first prompt Prompt_name_max - returns full date (April 01, 2011) for the second entry Prompt_name_max_label – returns results in date9. format. (01Apr2011) Prompt_name_max_rel – the relative term the user selected for the second prompt
Time	Enter the value by using a clock prompt. Users can also select a relative term (such as **Previous Hour** or **Next Minute**) Time Prompt Example Time: 08:18:21 PM OK	Prompt_name - returns the entry in time. format (16.30.22) Prompt_name_label - returns the entry in timeampm11. format (4:30:22 PM) Prompt_name_rel – the relative term that the user selected

Prompt Type	Description	Available Macros
Time Range	Provides two time entry boxes with the same options as the Time prompt. Time Range prompts are validated for proper range (small to large). Also, a range type is allowed for quick entry in the two time entry boxes. Time Range Prompt Example	Prompt_name_min – returns the entry in time. format (16.30.22) for the first prompt Prompt_name_min_label - returns the entry in timeampm11. format (4:30:22 PM) for the first prompt Prompt_name_min_rel – the relative term that the user selected for the first prompt Prompt_name_max - returns in time. format (18.30.22) for the second entry Prompt_name_max_label – returns the entry in timeampm11. format (6:30:22 PM) for the second prompt Prompt_name_max_rel – the relative term that the user selected for the second prompt
Timestamp	Note that the value is almost in the datetime19. format. You can convert to datetime19. by completing a code step to convert space to : using the TRANWRD function. Here is an example in a DATA _NULL_ step: `input(tranwrd("&prompt_name", " ", ":"), datetime19.);` Also, a range type is allowed for quick entry. Timestamp Prompt Example	Prompt_name – returns the entry in time. format (01Apr2011 16:30:22) Prompt_name_label - returns the entry in timeampm11. format (4:30:22 PM) Prompt_name_rel – the relative term that the user selected

Prompt Type	Description	Available Macros
Timestamp Range	Provides two timestamp entries. Once entered, the prompt automatically tests for a valid range selection prior to executing the project. Also, the same range types within the Timestamp prompt are provided to allow quick entry of values. Note: The conversion of the prompt to a valid datetime format is discussed in the description of the Timestamp prompt. Timestamp Range Prompt Example From: To: Current date and time Current date and time of previous year Current date and time of next year End of current hour End of previous hour End of next hour End of current minute End of previous minute	Prompt_name_min – returns the entry in time. format (01Apr2011 16:30:22) for the first prompt. Prompt_name_min_label - returns the entry in timeampm11. format (4:30:22 PM) for the first prompt Prompt_name_min_rel – the relative term the user selected for the first prompt Prompt_name_max - returns in time. format (18.30.22) for the second entry Prompt_name_max_label – returns the entry in timeampm11. format (6:30:22 PM) for the second prompt Prompt_name_max_rel – the relative term that the user selected for the second prompt
Data Source	Select a data source to use in the project. Data Source Prompt Example	Prompt_name – returns the metadata path of the data source selected Prompt_name_type – provides a numeric response for what the user selected so that the programmer can devise different actions 1 for tables 2 for OLAP cubes 4 for relational information maps 8 for OLAP information cubes

Prompt Type	Description	Available Macros
Data Source Item	Select a variable in a data source to use in the project. Data Source Item Prompt Example	Prompt_name – the variable Prompt_name_path – the metadata path for the data table selected Prompt_name_type - provides a numeric response for what the user selected so that the programmer can devise different actions 1 for tables 2 for OLAP cubes 4 for relational information maps 8 for OLAP information cubes
File or Directory	Select a file or directory structure to analyze in the project. The prompt is identical to the Data Source and Data Source Item prompts. However, the **Browse** button allows you to navigate through the Workspace Server file structure.	Prompt_name –physical file path of the file or directory selected Prompt_name_server – name of server that contains the files or the directories the user can select from. Server is defined by the prompt creator.
Color	Select a color from a color palette. The result is returned in hexadecimal format. Color Prompt Example	Prompt_name – If the user selects red, it returns as cxff0000 This prompt can be used to customize graphical output.

Prompt Type	Description	Available Macros
Data Library	Browse through the metadata to select the library to use in the code. Data Library Example Library: /Projects/Candy/Data/BID(Library) Libref: bid Browse...	Prompt_name – LIBREF Prompt_name_path – the metadata data library location
Variable	Select from a list of variables that the prompt creator defines. Variable names are of only one type. Option to allow users to manually type a variable name is also available in the Edit Prompt window. Note that only one entry is returned to the code.	Prompt_name – variable selected An example of when to use this prompt is if you have multiple date fields (such as open date, close date, on hold date, pending date) that the user might want to search through. A combination of the Variable prompt and the Date Range prompt would be useful to generate the query.

4.8 SAS Administrator Tasks

Using SAS Management Console, the SAS administrator can set responsibilities and make system-wide changes that assist all users.

4.8.1 Access Requirements

From SAS Management Console, the following items are required for users to use and for developers to share prompts.

1. Developers need Write Member Metadata access to create shared prompts in a metadata folder.

2. Users require Read Metadata access to the metadata folder that contains the prompts.

3. Users require Read access to any source data the prompt uses to dynamically populate the available values.

4.8.2 Allowing Limited Access to SAS Management Console

Since you can only share prompts through the SAS Management Console interface, it might be necessary to grant business users access to SAS Management Console on a limited basis. To limit the functionality for these users, creating a new role is recommended. Included are the steps.

1. From User Manager plug-in, select **New** -> **Role** from the **Actions** menu.

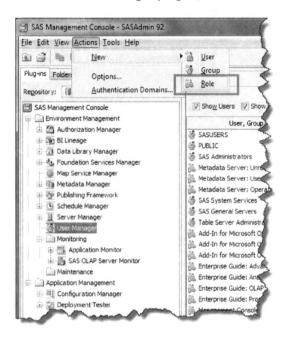

2. On the General tab, we named this role 'Sharing Prompts Role.'

3. Move to the Capabilities tab and select Folder View as highlighted in the following figure.

4. The SASUSER group is defined to SAS Management Console: Content Management Properties role by default. This will need to be changed as these permissions and capabilities are inherited by all SAS users. Open up the SASUSER group, on the Roles tab, select Management Console: Content Management and click one of the arrows to deselect this capability for users.

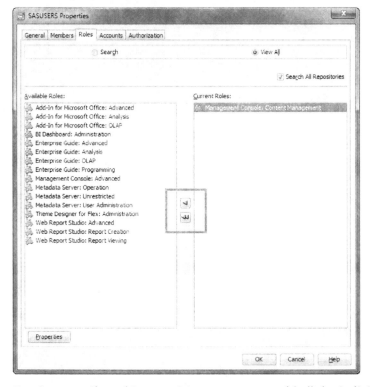

5. Create a Shared Prompt Manager group, add all the individuals who need this access, and on the Groups and Roles tab select the newly created 'Sharing Prompts Role'.

6. After installing SAS Management Console for these users, they will now have the ability to log in and see the SAS Folders tab and create their own Shared Prompt Manager stored process.

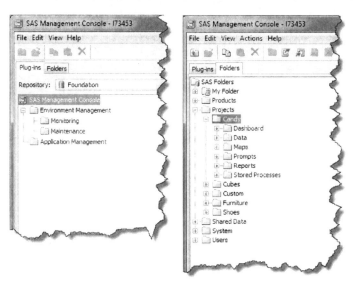

Chapter 5

SAS OLAP Cube Studio

Building Different Views of Your Data

Chapter 5

SAS OLAP Cube Studio

Building Different Views of Your Data

SAS OLAP Cube Studio allows you to easily and quickly transform your data into an online analytical processing (OLAP) structure. Data stored in an OLAP structure, which is called a cube, can be accessed from multiple SAS BI clients, including SAS Enterprise Guide, SAS Web Report Studio, and SAS BI Dashboard, to complete analysis, build reports, and even build other data structures.

An OLAP cube works particularly well with large data sets by reducing the time needed to access and display data. Performance is enhanced because data is summarized into measures and predefined into categories. Users are able to review the data from multiple angles so they can quickly answer questions without waiting for data summarizations or running various reports on the same source. The data can also be secured at levels within the cube, so one source can service many different users in the organization.

SAS OLAP Cube Studio guides you through the cube-building and maintenance process. This chapter explains how to use SAS OLAP Cube Studio to build a simple cube and introduces the features of the tool.

5.1 Getting Started

This topic provides an overview of the tool capabilities and what you need to get started.

5.1.1 Quick Tour

When you open SAS OLAP Cube Studio, the main window appears. From this window you can build, modify, and manage OLAP cubes. The two main areas you can use to navigate the metadata are the **Folder** and **Inventory** tabs. The following figure shows the default configuration for the application. Your system can display different information depending on the security structures and prior steps taken to organize metadata.

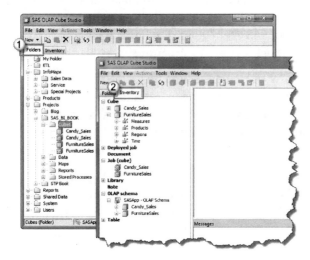

Figure 5.1-1. SAS OLAP Cube Studio main window

1	The **Folders** tab provides a navigation tree for the metadata.
2	The **Inventory** tab provides a navigation tree for the metadata objects. The **Cube** folder shows all cubes that you can access. You can see the Candy_Sales cube (this is the same cube as shown in **Projects** folder) and some other cubes that are not stored in the same location as the Candy_Sales cube. This tab provides a quick way to access all objects.

5.1.2 Prerequisites

Prior to using the application at your site, you must have the following:

- SAS OLAP Cube Studio application installed on your desktop
- Connection profile to the SAS Metadata Server
- Permissions to an OLAP schema
- SAS folder in the SAS Metadata Server to access and store the OLAP cube

Contact your SAS site administrator for more information or additional assistance.

Note: All examples within this chapter use the SAS OLAP Cube Studio; however, SAS Data Integration Studio provides the same graphical interface.

5.2 Understanding OLAP Cubes

OLAP cubes dramatically improve the ability to query large data structures. For small data structures, filtering on one or more categories and summarizing columns might take a few seconds, but for large tables that contain millions of records doing the same action could take several minutes. The other advantage to using cubes is that you can explore measures from a variety of viewpoints quickly.

The following topic explains how cubes are different from relational data structures and the terms used to describe a cube.

5.2.1 Understanding the Cube Structure

By now, you might be wondering how OLAP cubes are different from other data structures. If you have been working with reports and data, you are probably more familiar with relational or transactional data structures, where the data appears like a spreadsheet. The spreadsheet columns can be categorized, filtered, and counted. This is called transactional data because each row of the data represents a single transaction.

In the following figure, each row has the sales amounts summarized by category for the sales amount. You can also see how the data is organized; for example, ❶ State is a category for Country and ❷ PRODTYPE is segmented by Product. From the ❸ Date column, you could easily extract data for the year, quarter, or month.

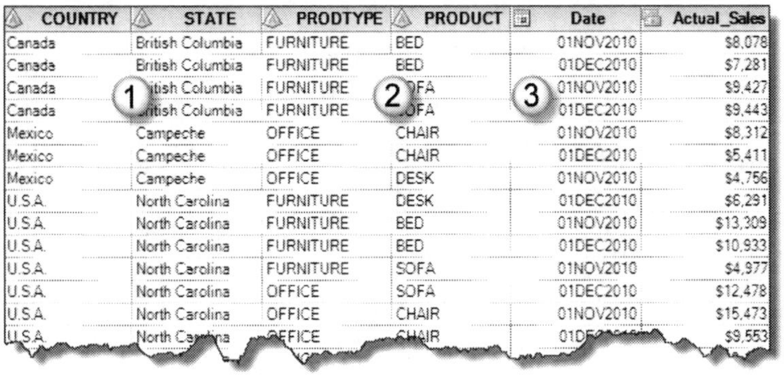

COUNTRY	STATE	PRODTYPE	PRODUCT	Date	Actual_Sales
Canada	British Columbia	FURNITURE	BED	01NOV2010	$8,078
Canada	British Columbia	FURNITURE	BED	01DEC2010	$7,281
Canada	...itish Columbia	FURNITURE	...FA	01NOV2010	$9,427
Canada	...itish Columbia	FURNITURE	...OFA	01DEC2010	$9,443
Mexico	Campeche	OFFICE	CHAIR	01NOV2010	$8,312
Mexico	Campeche	OFFICE	CHAIR	01DEC2010	$5,411
Mexico	Campeche	OFFICE	DESK	01NOV2010	$4,756
U.S.A.	North Carolina	FURNITURE	DESK	01DEC2010	$6,291
U.S.A.	North Carolina	FURNITURE	BED	01NOV2010	$13,309
U.S.A.	North Carolina	FURNITURE	BED	01DEC2010	$10,933
U.S.A.	North Carolina	FURNITURE	SOFA	01NOV2010	$4,977
U.S.A.	North Carolina	OFFICE	SOFA	01DEC2010	$12,478
U.S.A.	North Carolina	OFFICE	CHAIR	01NOV2010	$15,473
U.S.A.	North Ca...na	...FFICE	...AIR	01D...	$9,553

Figure 5.2-1 Example of Relational Data Source

If you want to create a report for this data, there are several ways to organize the information, such as summarizing the sales as year-to-date amounts by product, geography, or some combination of all. However, it would take several different reports to display this data for analysis. If this data is actually a million-record data table, the report could take a while to display.

Many times users are not trying to determine how many sofas were sold on a certain day; rather, they want the data summarized by country, state, product, and date so they can slice and dice the data to find patterns and trends. An OLAP cube makes this possible by providing all of the likely paths through pre-calculated measures across the data.

Relationships between data, such as Country and State; Product Type and Product; or Year, Quarter, and Month can be logically grouped into *dimensions*. Each grouping contains defined *hierarchies* that establish a path for users to drill into for more details. For example, a time dimension containing data for Year, Quarter, Month, and Day might contain two hierarchies. One hierarchy takes users from Year to Month to Day and the second hierarchy allows them to drill from Year to Quarter to Month. Each of the individual data columns used to compose the hierarchy is termed a *level*.

The following figure represents a cube structure with each side corresponding to a dimension and hierarchy. The measure within the cube is Actual Sales. Turning the cube or drilling into the hierarchy the measure Actual Sales will have a different sum based on your current view.

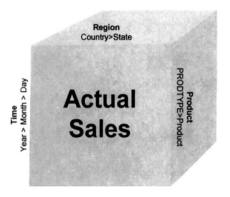

Figure 5.2-2 Cube with multiple dimensions

In the following figure, you can see how a cube looks in the SAS Enterprise Guide OLAP Viewer. You can see the Region and Product dimensions for the sales amount as sum and average. In Section 5.4, "Creating Your First OLAP Cube," you will learn how to create this cube.

Year		BED		SOFA		CHAIR	
Product		BED		SOFA		CHAIR	
Measures		Sum of ACTUAL	Average ACTUAL	Sum of ACTUAL	Average ACTUAL	Sum of ACTUAL	Average ACTUAL
Country	State/Province						
Canada		$224,162.00	$778.34	$233,436.00	$810.54	$202,283.00	$702.37
⊟⊞ Canada	British Columbia	$53,587.00	$744.26	$61,295.00	$851.32	$53,486.00	$742.86
	Ontario	$52,845.00	$733.96	$55,350.00	$768.75	$43,332.00	$601.83
	Quebec	$58,089.00	$806.79	$60,216.00	$836.33	$55,405.00	$769.51
	Saskatchewan	$59,641.00	$828.35	$56,575.00	$785.76	$50,060.00	$695.28
Mexico		$151,548.00	$526.21	$142,341.00	$494.24	$141,300.00	$490.63
⊟⊞ Mexico	Baja California Norte	$35,195.00	$488.82	$37,341.00	$518.63	$33,582.00	$466.42
	Campeche	$37,807.00	$525.10	$32,623.00	$453.10	$35,489.00	$492.90
	Michoacan	$39,558.00	$549.42	$39,405.00	$547.29	$35,331.00	$490.71
	Nuevo Leon	$38,988.00	$541.50	$32,972.00	$457.94	$36,898.00	$512.47
U.S.A.		$672,097.52	$777.89	$674,868.84	$781.10	$641,112.82	$742.03
	California	$121,759.10	$1,691.10	$123,069.80	$1,709.30	$112,332.60	$1,560.18
	Colorado	$78,453.00	$1,089.63	$64,973.00	$902.40	$68,446.00	$950.64
	Florida	$69,644.00	$967.28	$75,049.00	$1,042.35	$73,560.00	$1,021.94

Figure 5.2-3 Example of a cube in the OLAP Viewer

5.2.2 Understanding the Cube Terms

There are five cube terms to learn: *dimensions, hierarchies, levels, members,* and *measures.* In the following figure, a cross-tabular report is shown using cube data.

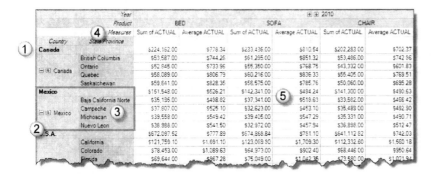

Figure 5.2-4 Cube structure layout example

This table defines each cube term listed in the previous figure and provides an example data item or structure.

	Term	Definition	Examples
1	Dimensions	Provides a grouping of natural elements that helps answer the key questions who, what, where, when, and how. In the example, the dimensions are country and product.	<ul><li>Location</li><li>Time</li><li>Product</li><li>Customer</li></ul>
2	Hierarchies	Identifies the path to move along for more details and summarizes the dimensions. This can be thought of as a parent-child relationship. For example, in the figure, Country is the parent to State/Province. At a minimum, one hierarchy is defined within each dimension. **Note:** The total number of hierarchies used across all dimensions cannot exceed 128.	<ul><li>Country -> State -> City</li><li>Region -> Subsidiary</li><li>Year -> Quarter -> Month</li><li>Year -> Month -> Day</li><li>Product Type -> Model -> Model Number</li><li>Alphabet -> Customer Last Name -> First Name</li></ul>
3	Levels	An individual column that is used to make up a hierarchy. Levels in the above example include Country, State/Province, and Product. **Note:** Within a hierarchy, up to 19 levels can be used. The total number of levels that can be used within a cube is 256.	<ul><li>Country</li><li>Region</li><li>Subsidiary</li><li>Product</li><li>Year</li><li>Month</li></ul>

	Term	Definition	Examples
4	Measures	Identifies the statistics available for each dimension. In the example, Sum of Actual Sales and Average of Actual Sales are measures. You can also create custom measures based on other measures. For example, percentages or ratios. **Note:** A maximum of 1024 measures per cube is possible.	• Count • Sum • Average • Standard deviation • Standard error of mean • Upper/Lower confidence limit • Count of missing values • NUNIQUE
5	Members	Represents an individual data element. You can use members in PROC OLAP code to create advanced measurements. **Note:** The maximum number of members is 2^{32} per hierarchy.	• [All_region_sub].[Asia].[Seoul] • [2011].[01].[10] • [Hardware].[Mouse].[87654] • [Doe].[John]

5.3 Creating OLAP Cubes

There are three steps to follow when creating OLAP cubes:

1. Preparing the source data

2. Designing the structure (dimensions, hierarchy layout, and measures)

3. Creating and maintaining the cube

5.3.1 Preparing the Source Data

Before you can build a cube, you must have a data source. A data source can be a single SAS data set or a group of star schema data sets, which are joined when the cube is created. Your choice for the data source depends on several factors, such as available tables, amount of data, and user requirements.

SAS OLAP Cube Studio allows you to use one of three different data sources: detail table, fully summarized table, or star schema tables.

5.3.1.1 Detail Tables

This data source closely resembles the spreadsheet structure: it contains information for every measure and member necessary to generate the cube structure. In the following example, you can see an example extract of a detail table. This data represents individual transactions for each product for a particular month. SAS OLAP Cube Studio summarizes these records into the defined dimensions.

Country	State	Type	Product	Date	Actual Sales
Canada	Quebec	Office	Chair	01FEB11	10,000
Canada	Quebec	Office	Chair	15FEB11	12,000
Mexico	BAJA	Office	Chair	01FEB11	20,000
Mexico	BAJA	Office	Chair	15FEB11	12,500
USA	CA	Office	Chair	01FEB11	74,000
USA	CA	Office	Chair	15FEB11	75,500

Table 5.3-1 Example of detail table data source

5.3.1.2 Fully Summarized Table

This data source is already summarized to the lowest level for all dimensions. The following figure shows an extract of the data in Table 5.3-1 after it has been summarized. The actual sales have been totaled by each dimension, resulting in the February sales.

Country	State	Type	Product	Month	Actual Sales
Canada	Quebec	Office	Chair	FEB11	22,000
Mexico	BAJA	Office	Chair	FEB11	32,500
USA	CA	Office	Chair	FEB11	149,500

Table 5.3-2 Example of a fully summarized data source

When the data is summarized to its lowest level of all analyzed variables, it is called an *n-way summary*. You can use a SAS procedure called PROC SUMMARY to create summarized data before using SAS OLAP Cube Studio. Following is an example of PROC SUMMARY that creates the fully summarized table.

```
/*===========================================================*/
/*Create an NWAY summary for all levels in the data set */
         proc summary data=SASHELP.PRDSAL3 nway;
         class country state prodtype product date;
         var actual_sales;
         output out=mylib.class_sum sum(actual_sales)=actual_sales;
         run;
/*=============================================================*/
```

Program 5.3-1 Sample summary procedure code for PRDSAL3

5.3.1.3 Star Schema Tables

Star schema data sources consist of multiple dimension tables and one additional table known as the fact table. Fact tables are joined to the dimension tables by key variables. In the following figure, a fact table is in the center with four dimension tables surrounding it, resembling a star. The fact table contains the numeric or aggregate values for each dimension, and the dimension tables contain the data categories. For example, in this figure, the fact table has several key variables, such as Prod_ID, Region_ID, and so on. For each key variable, there is a dimension table with a matching key.

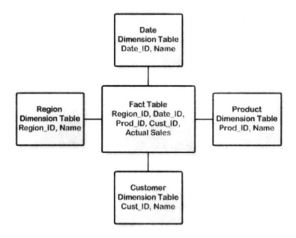

Figure 5.3-1 Star schema data source

In the following figure, you can see an example of how the raw data in each table appears. The fact table ❶ lists the ProdID key variable with the value of 7. This correlates to the ProdID in the Product dimension table❷. The Product dimension table has a hierarchy of Category > Subcategory > Product. SAS OLAP Cube Studio automatically creates the join when you define the dimension.

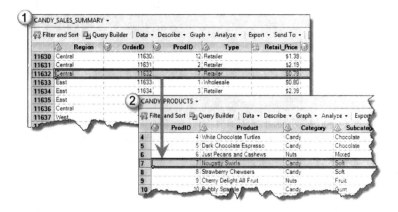

Figure 5.3-2 Linking from a Fact table to a Dimension table

For larger tables with many dimensions, it is easier to manage the data when it is organized in this structure. The benefit of using a star schema is to speed data retrieval and have the data formatted in a way that is easy to understand and maintain.

The snowflake schema is a variation of the star schema where two or more dimension tables are joined together before joining with the fact table. An example of the snowflake schema is a product dimension table that contains IDs that link to product type and group dimension tables, while the fact table contains an ID linking only to product dimension. Snowflake schemas are not directly supported by SAS OLAP; conversion of the dimension table joins into a single table (or view) is required before beginning the OLAP definition.

 In order for SAS OLAP to retrieve snowflake schema's additional dimension tables, the tables must first be converted into a single dimension table or view.

5.3.2 Designing a Cube Structure

Designing a cube is somewhat of an art. After building an initial cube, you might have to redesign it several times before it is ready for production. Preparing the data, understanding the requirements, and planning the dimensions and measures will increase your success.

5.3.2.1 Organizing Your Data

After learning the OLAP terms, you might already have some ideas about the cube structure. The logical dimensions and hierarchies, such as time (year, month, day) or geographical (country, state, city) and logical measures, such as sum of sales and count of units, will quickly fall into place.

When creating the cube, it is helpful to begin with the end in mind. The first thing to consider is how end users will interact with the cube when it is used in a report and what they are trying to measure. For instance, department managers are interested in how close, in dollars, they are to exceeding their budget. For the cube to be useful for the end user, the data must be logical and easy to navigate. Users might quickly abandon the cube if the path or level they need is missing.

Before starting SAS OLAP Cube Studio, you must prepare your data. Consider the following questions when designing the cube:

- Dimensions
  - What paths will end users explore?
  - What questions are the users asking about the data?
  - What natural hierarchies and levels exist in my data, such as Time and Location?
  - What hierarchies do I need to create?
    - Product lines might need data from several tables moved to one
    - Departments structures might not exist in a central area or be maintained
- Measures
  - How does the data need to be calculated?
    - Widgets sold
    - Mean time to failure
    - Average customer calls per month
  - Do I have the needed data to generate the calculations?

5.3.3 Creating and Maintaining OLAP Cubes

OLAP cubes can be created in one of two ways: writing the code in SAS Enterprise Guide using the OLAP procedure or using SAS OLAP Cube Studio.

In some sense, the cube design process never ends. As the end users employ the data for analysis and reporting, they will discover additional needed dimensions and measures. Cube maintenance includes tasks such as scheduling cube refreshing, reducing storage space, or improving performance of queries for end users. In Section 5.7, "Maintaining an OLAP Cube," you will learn more about cube maintenance.

 SAS also supports access to its OLAP technology using multidimensional expressions (MDX), which allows for custom queries and measurements using functions similar to those found in other vendor applications, such as Microsoft Analysis Cubes, Oracle, and SAP.

5.4 Creating Your First OLAP Cube

This topic demonstrates how to use the SAS OLAP Cube Studio wizard to create a cube for a fictitious furniture company. The data is detailed in one table and the end result is shown in Figure. This simple cube has two dimensions and several measures.

Note: The OLAP cube created in this example is used to create a report in Chapter 7, "SAS Web Report Studio."

5.4.1 Creating an OLAP Cube from Detailed Data

Use the following steps to create a cube:

1. Open SAS OLAP Cube Studio and select **New>Cube** to start the Cube Designer wizard.

2. Complete the fields in the initial page Cube Designer – General and select the **Next** button to continue.

    Note: The numbers in the table correspond to the following figure.

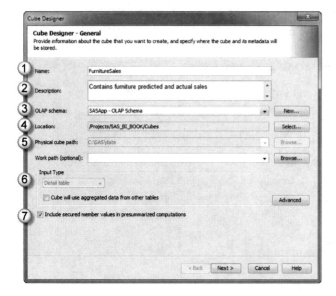

	Field Name	Description
1	**Name**	The cube name can be up to 32 characters long. Do not include spaces within the name field. Certain functions do not act as expected when spaces are embedded within the name.
2	**Description**	Enter a description of what the cube contains. This information can be useful for users accessing the cube from other locations.
3	**OLAP schema**	Displays the list of available OLAP schemas defined within the metadata server. The OLAP schema defines to the metadata server which SAS OLAP Server runs the cube. The default OLAP schema is **SASApp—OLAP Schema**.
4	**Location**	Choose a metadata folder location where appropriate security is applied. This is where the users can find the cube.
5	**Physical cube path**	Select a physical path on the server to store the cube.
6	**Input Type**	Select the input type that matches your data source. There are three choices: **Detail Table**, **Star Schema**, and **Fully Summarized Table**. Refer to Section 5.3.1, "Preparing the Source Data," for a more complete description.
7	**Include secured member values in presummarized computations**	*Optional* If member-level security is applied and you want to enforce it throughout views, ensure that this check box is blank. Refer to section 5.8.2, "Member-Level Security," for more detail.

3. Select your data sources in this step by defining the input and drill-through data tables. Depending on the input type selected in the first step, these windows can be slightly different.

Note: If you are creating a cube from with a star schema data source, refer to Section 5.5, "Creating a Cube from Star Schema Tables," for more information.

The first window requests the Input table, which contains all levels and measures required to build the OLAP cube. The second window requests the Detail table, which is used when users are allowed to drill through to detail on a report. If the data source does not appear in the available tables, it might not be registered with the metadata server. If you have Write Metadata Access to a particular folder, you can define new tables on this window by selecting the **Register Table** button.

 Click the **View Data** button to see a preview of the source data to ensure that you are selecting the correct table.

a.   In the Cube Designer – Input page, select the input data source. Move the data source to the right side. In the following figure, PRDSAL3CUBE was selected and displays in the **Selected table** area. Click the **Next** button to continue.

b.   In the Cube Designer – Drill-Through page, select the data source that is used for the drill-through. Typically, when using the Detail Input type, the input and drill-through uses the same data source. Click **Next** to continue.

   In fully summarized tables, you can use the same table or the source table that was used when creating the fully summarized table. When using a different table as the detail source, you must ensure that each variable in the cube design is represented in the table.

4.   In the Cube Designer – Dimensions page, you can setup the dimensions. Click the **Add** button to create a dimension.

For this example, this is a simple dimension called Region comprised of Country and State. The raw data is shown in Figure 5.2-1, Example of Relational Data Source.

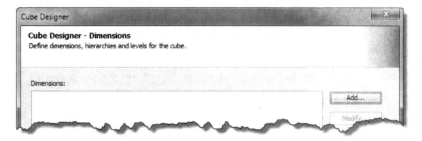

c. Complete the fields in the Dimension Designer-General window and click **Next** to continue.

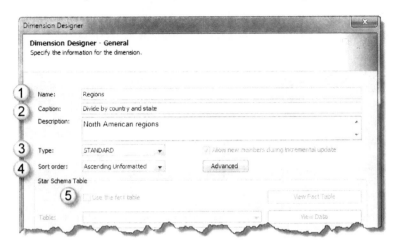

	Field Name	Description
1	**Name**	The name of the dimension cannot be the same as the name of any levels.
		If you have a variable called Region in the data table and you name the dimension Region, an error message is generated. In the example, the dimension is called Region.
2	**Caption Description**	Use the **Description** field to explain what levels are included in the dimension. This field is optional.
		The user sees this value when viewing the cube directly from BI clients such as SAS Add-In for Microsoft Office and SAS Enterprise Guide. It is a good practice to make the description as descriptive as possible.
3	**Type**	You can select three types: **Standard, Time,** and **GEO**.
		Most of the time the type is Standard. Refer to Section 5.6.1, "Adding Time and GEO Dimensions," for more information about how to use the other types.
4	**Sort order**	Affects how the data is displayed when directly accessed from tools such as SAS Add-In for Microsoft Office and SAS Enterprise Guide.
		![icon] Sort order is not maintained when a cube is accessed through an information map.
5	**Star Schema Table**	When dimension data is contained within the fact table itself you should select this option. This is used only when working with data arranged as a star schema.
		Refer to Section 5.5, "Creating a Cube from Star Schema Tables," for information on adding dimensions for star schema data.

d. In the Dimension Designer – Level page, you can select the levels that belong to the dimension. The first dimension, called Regional Stores, has two levels, Country and State. To define the Levels, click the **Add** button.

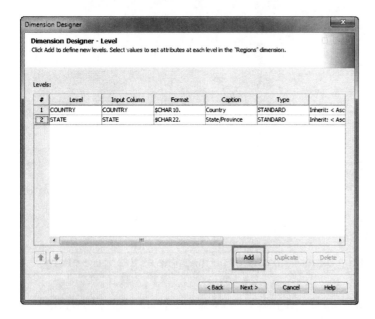

e. In the Add Levels window, select the variables you want to add. Click **Next** to continue.

 If there will be only one hierarchy, you can choose the levels in the order they are used in the hierarchy. This step saves time during hierarchy definition.

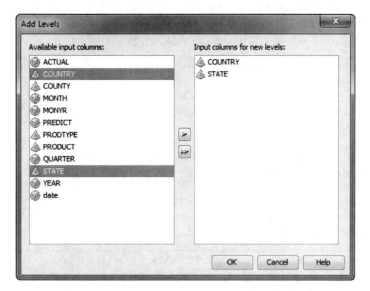

Chapter 5: SAS OLAP Cube Studio   147

f. To set the hierarchy order for the dimension, you must add a hierarchy. Click the **Add** button to create a hierarchy.

Select the **Finish** button for SAS to automatically create the hierarchy and skip the next step. For this step to work correctly, select the data fields for your dimension in the desired hierarchal order.

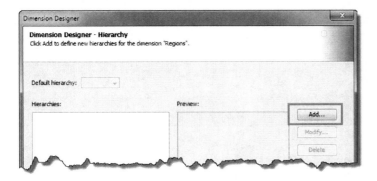

g. In the **Available input columns** area, drag the fields to the selected area in the correct order. Select **OK** to continue. Then click the **Finish** button to continue creating the cube.

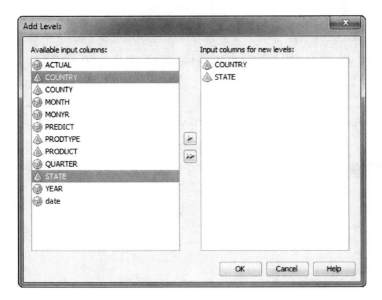

5. In this step, you select the measures used in the cube. You must select at least one available measure to move forward. These measures are calculated during the cube building process; therefore, certain calculations (such as a ratio) are not available at this point.

 Calculated members such as a ratio are defined and stored within the metadata and calculated at run time, no matter the cube type.

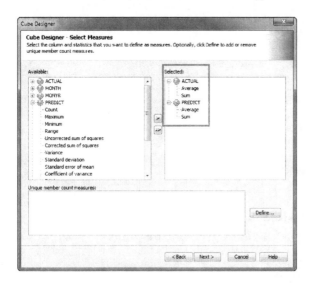

In the lower half of this window, you have the ability to define measures based on unique member counts. An example is when you need the **Number of Customers** as a measure on your report, but your data table contains the customer name.

6. In the Cube Designer - Measure Details page, you can set the default measure and adjust the labels and formats for measures.

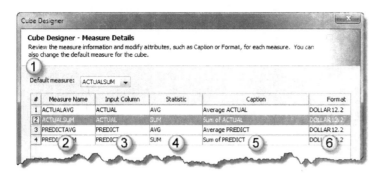

	Field Name	Description
1	**Default measure**	Define the default measure users see when initially opening the cube directly in SAS Add-In for Microsoft Office or SAS Enterprise Guide. Select the one that seems more appropriate. In this case, sum of sales is chosen.
2	**Measure Name**	This name is used in backend queries; these names must not have spaces.

Chapter 5: SAS OLAP Cube Studio   149

	Field Name	Description
3	**Input Column**	The source table column used to derive the measure.
4	**Statistic**	The statistic function used to summarize the data.
5	**Caption**	Name of the measure that is displayed for users. You can modify this name so it is more user-friendly if necessary.
6	**Format**	Format of the measure that is displayed to users. You can modify this format to match the statistic. For instance, the DOLLAR12 option causes the dollar sign ($) to display in front of the value.

7. In simple OLAP structures, no member properties are required for an OLAP cube. This step is optional and you do not need to change the default settings selections in this window. Select **Next** to continue.

    Member properties are attributes of dimension members that provide additional information to cube users. Member property information is usually not as significant as the levels and members within a dimension, and therefore does not qualify as a level or member. However, it can have additional analytical value that can be useful at query time.

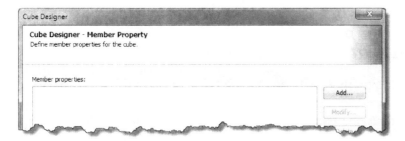

8. You can add aggregations to the cube during the creation process. This step is optional and you do not need to change the default settings selections at this window. Click **Next** to continue.

    For more information about adding aggregations or using aggregations to improve cube performance, refer to Section 5.7.2, "Adding Custom Aggregations with the Aggregation Tuning Module."

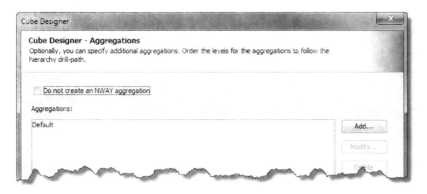

9. When the wizard is complete, you can review the cube information. To build the cube, select the first radio button. If you have created this cube before, this button deletes and re-creates any existing cubes.

 To save the code for use in SAS Enterprise Guide, select the **Export Code** button.

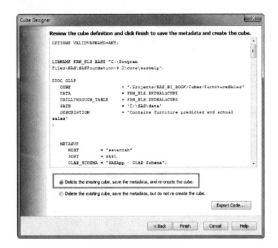

5.5 Creating a Cube from Star Schema Tables

When creating a cube with star schema tables, there are two times in the cube creation process where the process varies from how the other cube types are created. The first is when selecting the data and the second occurs when creating the dimensions. For a description of star schema tables, refer to Section 5.3.1.3, "Star Schema Tables."

5.5.1 Defining Star Schema Tables

When building a star schema cube, you need to specify the input fact table, dimension tables, and a drill-through table. In the Cube Designer - Input page, provide the fact table ❶ for the cube, shown in the following figure. The fact table contains the join information and measures. In the Dimension Tables page, add all the required dimension tables ❷ to the **Selected tables** area.

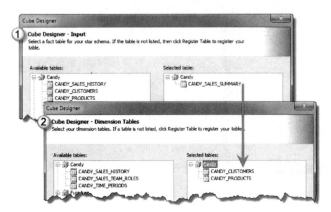

Figure 5.1-1 Choosing data sets for star schema data sources

In the next step, you are prompted to provide the drill-through table. The drill-through table in this scenario requires a view either with all the joins defined or a full detail table.

 With star schema structures, an SQL view must be developed and registered separately for drill-through to detail functionality to work. Users can create this join in SAS Enterprise Guide or SAS Data Integration Studio and update library metadata with the table structure.

5.5.2 Adding Dimensions with Star Schema Tables

When adding dimensions using star schema data sources, the Dimension Designer interface provides an area to define the relationship between the fact table and the dimension tables. In the dimension table, the hierarchy is Product > Category > Subcategory. The common variable ProdID joins the two tables when you build the dimension. Refer to Section 5.5, "Creating a Cube from Star Schema Tables," for an example of how star schema tables are joined.

To create a dimension from a star schema table, do the following:

1.   At the Cube Designer – Dimensions page, select the **Add** button to start a new dimension.

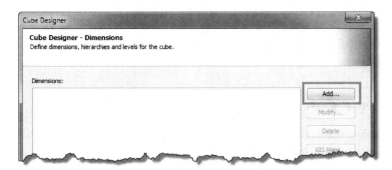

2.   In the Dimension Designer – General page, complete the **Star Schema Table** area, and then continue with the dimension creation.

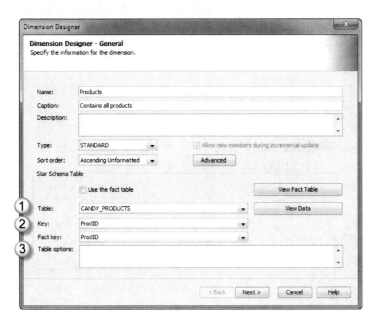

The check box to **Use the fact table** should be selected on the Cube Designer- Input table page when all the levels for the dimension reside in the fact table.

	Field Name	Description
1	Table	Select the dimension table with the key variable.
2	Key/Fact key	Values in these drop-down fields establish the relationship between the tables. In this example, ProdID is selected, but the variables do not have to have the same name. **Key** – Corresponds to a column in the dimension table. **Fact key** - Represents the fact table column.
3	Table options	You can set any data set options, such as a WHERE clause, for the selected dimension table.

5.6 Enhancing the Cube

SAS OLAP Cube Studio makes enhancing the cube easy. You can create time and location dimensions, address missing data, and create custom measurements.

5.6.1 Adding Time and GEO Dimensions

There are two special dimension types available when creating a cube: time and GEO. These dimension types have special properties that make creating the cube easier.

5.6.1.1 Adding Time Dimensions

When adding a time-specific dimension such as Year > Quarter > Month, you can take advantage of the built-in time hierarchies. After selecting Time as the dimension type, the **Add supplied time hierarchies** choice becomes available on the Dimension Designer – Level page, as shown in the following figure.

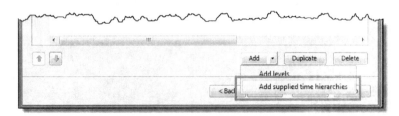

Figure 5.6-1 Add time dimensions

The **Add Supplied** window appears and you can choose a supplied time hierarchy structure. These supplied time hierarchies help build the dimension and auto-populate the levels and some level properties for the cube, reducing the time required to complete these windows.

 The value in the **Input column** field must be a date or date/time value for the hierarchy to work properly.

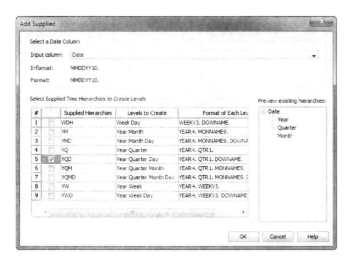

Figure 5.6-2 Using Add Supplied window

Time-series functions including rolling totals, rolling averages, and parallel period comparisons are enabled when a time dimension is defined, providing another benefit of using this dimension type.

5.6.1.2 Adding Geographic Locations with the ESRI Map Component

You can add geographic locations to your cube using the GEO dimension choice if you have an ArcGIS Server licensed at your organization. This is a separate product sold by Esri, not sold by SAS.

To use geographical locations, you must have the data set up properly. To avoid errors, data quality is extremely important, as there must be a match between the field ID in the map with the field ID in the cube input table. Within the Esri ArcMap, a layer should be defined for each level in the OLAP dimension. The ArcMap layer must contain the **Map Service** field ID information, along with the polygon details. An example of this mapping is shown in the following table.

OLAP Level	Cube Input Table Field ID Column	Map Service Layer	Map Service Field ID
State	StateFIP	TL_2009_State	StateFIPID
County	COFIP	TL_2009_County	CNTYIDFIP
City	PLCIDFP	TL_2009_Place	PLCIDFP

Table 5.6-1 GEO mapping data

To create the geographic location, do the following:

 Before adding a GEO dimension, create an Esri map using Esri ArcMap and create the Esri map service using the Esri ArcCatalog.

1. At the Dimension Designer – General page, select **GEO** in the **Type** drop-down box. The **GIS Maps** button on the Dimension Designer is enabled and available for use.

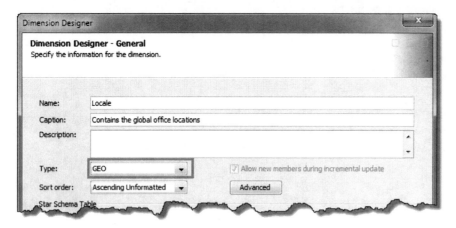

2. Select the **GIS Maps** button to open the GIS Maps Definition window. From here you must select a defined Esri Map Server and a running map service, and then connect the dimension levels to the map data.

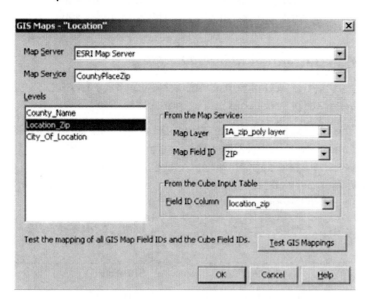

5.6.2 Working with Unbalanced or Ragged Hierarchies

Customarily, there is one value in each level through a hierarchy, such as for dates: Year, Month, Day; or 2011, June, 10. In situations where there are gaps in the hierarchy, there are several options that can be applied to individual dimensions or the entire cube. An example is an organizational chart where there are gaps in the levels of the hierarchy for different members, which is called a *ragged hierarchy*.

 Query performance is impacted when missing members are not appropriately defined.

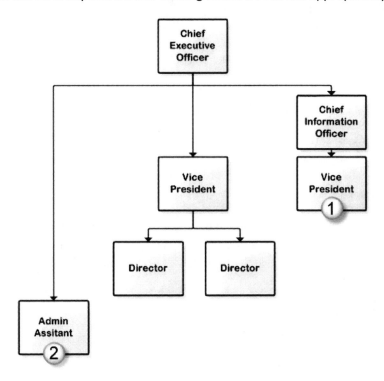

Figure 5.6-3 Ragged hierarchy example

If you consider the path to the Vice President ❶ under the Chief Information Officer, the member would look like the following:

[Organization].[All Organization].[Chief Executive Officer].[Chief Information Officer].[Vice President]

However, the member for the Administrative Assistant ❷ looks likes the following example:

[Organization].[All Organization].[Chief Executive Officer].[ ].[ ].[ ].[Administrative Assistant]

Because this member does not have the same structure, the missing levels are represented with empty brackets. You can adjust the settings for an entire OLAP cube level or for individual dimensions. When the settings are in place, the same member would look like the following example. In this member, the empty brackets are not in the path.

[Organization].[All Organization].[Chief Executive Officer].[Administrative Assistant]

5.6.2.1 Adjusting the Settings for Missing Data

To adjust the settings for the missing data, use the following hints.

Change settings for Entire Cube: To adjust the ragged hierarchy settings for the entire cube, select the **Advanced** button from the Cube Designer – General page. From the Advanced Cube Options window, click the check boxes to ensure that the missing data is handled appropriately.

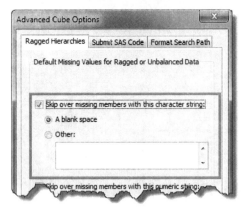

Figure 5.6-4 Change setting for entire cube

Change Settings for One Dimension: For a single dimension, select the **Advanced** button on the Dimension Designer – General page. In the Advanced Dimension Options window, you can choose to inherit the cube settings. You can also override the cube settings and make changes to the character and numeric values for the dimensions.

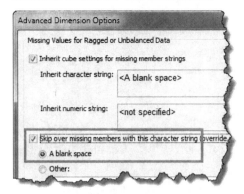

Figure 5.6-5 Change settings for one dimension

5.6.3 Defining Calculated Members for Custom Measurements

After creating the cube, you can add custom measurements. These measures are available for users from any interface and essentially appear the same as the measures you defined during the cube building process. The key difference is that these measures are not calculated and stored within the cube itself, but are only defined. When the user views the cube, the calculations are completed at run time and will change as the user manipulates the cube views.

There are three available starting points for a calculated member definition:

Simple GUI	• Sum
	• Difference
	• Ratio
	• Percent Increase
	• Percent Decrease
	• Distinct Count (note that the NUNIQUE function, which is defined during the measure definition in the cube building process, might perform better than Distinct Count)
Time	• Opening Balance
	• Closing Balance
	• Rolling Total
	• Average Over Time
	• Compare Parallel Periods
	• Compare Consecutive Periods
Custom	Opens a definition window to define the measure name and format. This allows you to use multidimensional expressions (MDX) code to create a custom measure. Microsoft Corporation initially developed the MDX reference code as part of the OLE DB for OLAP specification.

 Microsoft provides an MDX reference with both syntax and language documentation.

5.6.3.1 Creating a Custom Measurement

After developing the cube, you can open the Calculated Members window by right-clicking the OLAP cube and then selecting **Maintain > Calculated Members**. In the following figure, Sum of Actual was created within the Furniture Sales cube. The remaining three columns are calculated members using the Time Series option.

Calculated Members (Time Series) Examples

Quarter	Month	Sum Of Actual	Cons. Periods	Ytd Rolling Total	Rolling Avg (past 3 Periods)
	Jan	$348,337.79	$-938	$348,338	$348,765
1	Feb	$348,940.64	$603	$697,278	$348,851
	Mar	$358,416.44	$9,476	$1,055,695	$351,898
	Apr	$348,987.32	$-9,429	$1,404,682	$352,115
2	May	$348,398.97	$-588	$1,753,081	$351,934
	Jun	$344,124.62	$-4,274	$2,097,206	$347,170
	Jul	$357,014.66	$12,890	$2,454,220	$349,846
3	Aug	$354,791.61	$-2,223	$2,809,012	$351,977
	Sep	$343,344.02	$-11,448	$3,152,356	$351,717
	Oct	$359,212.29	$15,868	$3,511,568	$352,449
4	Nov	$360,393.21	$1,181	$3,871,962	$354,317
	Dec	$351,992.15	$-8,401	$4,223,954	$357,199

Figure 5.6-6 Creating custom measurements

To create a calculated member, do the following:

1.   Right-click the cube you want to have the new calculated member. Select **Maintain > Calculated Members**.

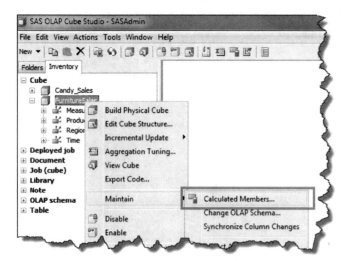

2.   From the Calculations window, select the **Add** button to create a new calculation.

3.   Select **Time Analysis Calculation** from the Calculation Type window.

4. In the Time Calculations page, select the **Average Over Time** radio button. For the **Formula**, select the existing measure, a time period, and the number of periods for the analysis. Select **Next** to continue.

    In the following figure, the new member is the average sales for the last three periods.

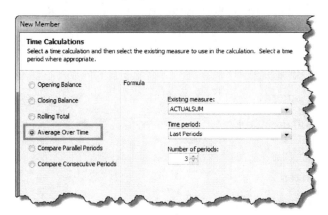

5. In the General page, type the name and select a format. Because the end user sees this name, try to be as descriptive as possible. You can also change the format. For this example, actual sales is a dollar value, so the format was changed to DOLLAR15. Click **Next** to save the measure.

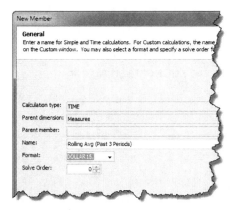

5.7 Maintaining an OLAP Cube

If a cube was not created properly or regularly maintained, you might notice that it takes longer to generate, requires more system resources, and still does not deliver the promised speed and ease of use for the end users. In this section, you will learn some techniques for keeping your cubes in top shape.

5.7.1 Variations in Data Storage Techniques

There are several flavors of cubes. The most confusing part of OLAP for many users is the differences in MOLAP, ROLAP, and HOLAP.

MOLAP	This is the default storage technique. This is where the summarization is calculated in advance and stored within the physical structure of the OLAP cube.
	This technique is best for performance because the aggregations are done ahead of time. All aggregations are stored on the server for quick retrieval. For large data sets, there might be sizing constraints for building the aggregations, making ROLAP a better choice.
ROLAP	The R represents relational, and gives users the ability to save the data outside of the OLAP cube in a flat or relational data set structure.
	This technique works well in these situations:

- Cube is built from fully summarized data tables

- Need to eliminate the OLAP update time is critical

- Physical space on the server is limited

	Performance for queries is dependent on the relational data tables. Users might find that this mechanism is much slower than MOLAP.
HOLAP	Hybrid OLAP offers some of both MOLAP and ROLAP storage techniques. Essentially, the structure is a ROLAP approach; however, the OLAP creator chooses to create several aggregations that are precalculated, so common queries and reports off this data generate results quickly.

When editing the cube or creating a new one, the Cube Designer – Aggregations page provides an option to turn off NWAY aggregation. Selecting this check box changes the cube data storage from MOLAP to either HOLAP or ROLAP.

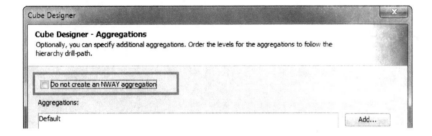

Figure 5.7-1 Setting the NWAY aggregation

5.7.2 Adding Custom Aggregations with the Aggregation Tuning Module

You can also create custom aggregations by selecting the **Add** button on the screen shown in Figure 5.7-1. However, with large or complex cubes, use the Aggregation Tuning dialog box from SAS OLAP Cube Studio. This mechanism analyzes log files or the data cardinality to assist with creating appropriate aggregations.

If users are complaining that OLAP-based reports are taking significant time to generate, consider adding custom aggregations to the cube. By default, OLAP cubes are stored with the NWAY aggregation only.

Use the Aggregation Tuning dialog box to analyze the OLAP logs and discover more dimensional cross joins to aggregate during the cube build process. Saved aggregations are stored within the OLAP definition and each time an OLAP cube is refreshed, these levels are recalculated and saved in the physical location with other aspects of the cube.

Within the dialog box there are three different mechanisms are available to add aggregations using this module: ARM Log, Cardinality, and Manual.

5.7.2.1 Creating the ARM Log

When using the ARM Log mechanism, the logging output must be increased to include the query detail. From SAS Management Console, complete the following steps to turn on logging:

1. Select the **Plug-ins** tab.

2. Expand **Server Manager > SASApp > SASApp - Logical OLAP Server**.

3. Right-click on **SASApp - OLAP Server** and select **Connect.** The application connects to the server and the right panel shows the connections.

4. Select the **Loggers** tab and locate the **Perf.ARM.OLAP_SERVER** logger. Select **Properties** to display the Perf.ARM.Olap_Server Properties window.

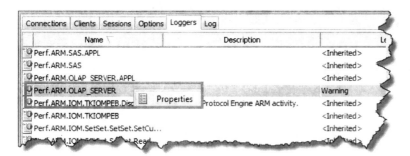

5. Click the **Assigned** radio button and select **Information** from the drop-down list. This allows the server to capture the cube usage. Click **OK** to exit.

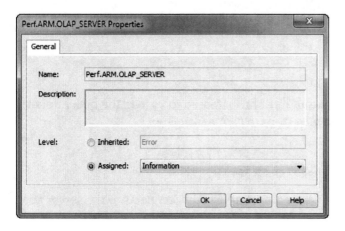

6. Use the cube in a manner that simulates actual usage for best results. For instance, if the cube is mainly used with SAS Web Report Studio, then run a report using that application. The log needs to capture a representative sample of the common OLAP cube uses. It is a good idea to use the cube in the other applications for additional data.

7. Once the log is completed and ready for use by the Aggregation Tuning function, you should repeat these steps to return the logging level to the **Inherited>Error** choice.

 The log files can grow exponentially and performance can be impacted when the logs are allowed to run without boundaries.

5.7.2.2 Analyzing the ARM Log

When the ARM log is ready for review, do the following to analyze the log.

1. In SAS OLAP Cube Studio, right-click the cube that you want to analyze and select **Aggregation Tuning** from the pop-up menu.

2. From the **Arm Log** tab, select the **Browse** button to point to the ARM Log created in the OLAP process.

 The typical location is within the configuration directory, such as:

`<configuration directory>\Lev1\SASApp\OLAPServer\Logs`

3. Select the radio button that creates the aggregation recommendations.

4. Select the **Analyze** button to define possible aggregations for inclusion within the OLAP cube.

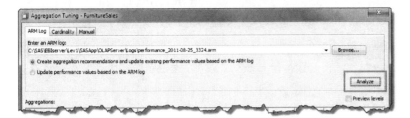

5. After aggregations are developed, you have the option to remove the aggregation from inclusion in the next OLAP build (using the **Delete** or **Drop** buttons).

6. Once completed, select the **Update Aggregations** button to create these aggregation tables for the OLAP cube to use.

The analysis on the **Cardinality** tab reviews the number of distinct members within each level of the cube; those with the highest cardinality are listed. Note that only 100 aggregations can be built at a time with this tool. However, after building a set of 100, you can add more.

In the following figure, you can see an example of recommendations based on the user interactions with the OLAP cube, as detailed in the ARM log created in the prior steps.

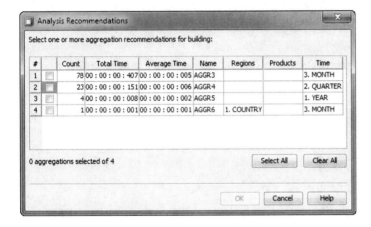

Figure 5.7-2 Example recommendations

5.7.3 Scheduling the OLAP Cube Refresh

Several mechanisms are available to schedule the OLAP cube refresh. From SAS Management Console, the OLAP job can be scheduled for routine refresh. Alternatively, you can export the short code and schedule the refresh using your native scheduler.

5.7.3.1 Using the Cube Job

Each time a cube is defined, a cube job is created automatically. The cube job provides metadata information, which can be used to deploy and schedule refreshes of the cube data or move a single cube from one system to another.

In the following figure, you can see where the jobs are located in each tool. In SAS Management Console, the cube jobs are located within the same metadata folder ❶ that the cube is stored in. In SAS OLAP Cube Studio, the **Inventory** tab has an expandable tab called **Job (cube)**❷.

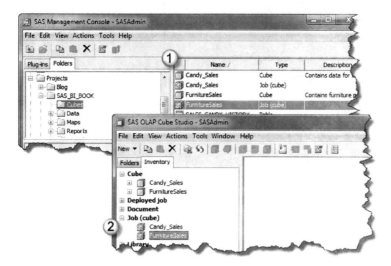

Figure 5.7-3 Cube job location

Before scheduling a job in the Schedule Manager plug-in for SAS Management Console, the job must be deployed for scheduling in SAS OLAP Cube Studio.

1. In SAS OLAP Cube Studio, right-click the job (cube) you want to change and select **Scheduling > Deploy** from the pop-up menu.

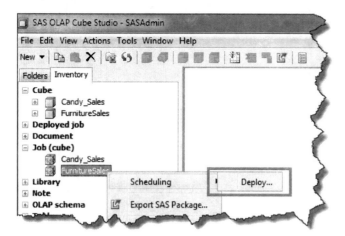

2. Define the job on the batch server, as shown in the following figure.

The deployment directory is the location for a .SAS file that you then schedule using the Schedule Manager in SAS Management Console.

 If you receive an error before this step, you must define a SAS DATA Step Batch Server within SAS Management Console. Refer to the SAS user documentation at support.sas.com for assistance.

3. From the Schedule Manager in SAS Management Console, select **New Flow** and schedule the deployed job for a routine refresh.

5.7.3.2 Using SAS Code

After creating the cube, you might want to make further changes to the code by using SAS Enterprise Guide, or provide the code to others. To export the code from SAS OLAP Cube Studio, select the cube and choose **Export Code**. In the Export Code window, select the **Long form** option and select the location to store the program.

Figure 5.7-4 Exporting the OLAP code

Open the newly exported SAS program and uncomment the first PROC OLAP step by surrounding the comment block top and bottom lines with /* */. Then a native scheduler can be used to set up the routine refresh.

5.7.4 Updating In-Place

If a user session is active and has the cube locked, it can block the cube from refreshing the data. Some organizations have programmed mechanisms to automatically stop and restart the SAS OLAP Server to force a disconnection of users before refreshing cubes. This requires limited cube refresh times, such as overnight or on the weekend. For international companies, this proved extremely problematic as the window that all users are off the system can be extremely limited. In SAS OLAP Cube Studio, you can refresh the cube with new data *in-place*.

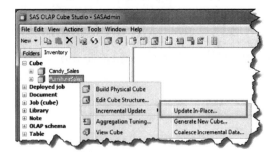

Figure 5.7-5 Select the Update In-Place menu choice

All existing connections to the cube remain intact while new connections use the new cube. Once all existing connections have been closed, the old cube is removed automatically. This action is done *in-place*, which means that it is invisible to the user community; the cube name, metadata location, and physical location remain the same.

The official switch between cubes can consume a significant amount of time if a large number of sessions access the cube; therefore, you can force an implementation of the refreshed data using the **Disable** and **Enable** menu elements.

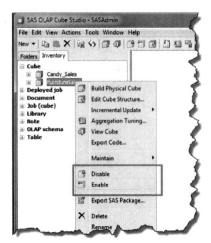

Figure 5.7-6 Select Enable menu choice

These steps can be completed programmatically and scheduled. The OLAP procedure has options ❶ (ADD_DATA, UPDATE_INPLACE, and UPDATE_DISPLAY_NAMES) that can be added to the code to perform an in-place update.

```
/* ================================================== */
PROC OLAP
 CUBE = "/Projects/Cubes/Candy_Sales"
 DATA = candy.CANDY_SALES_SUMMARY

 ❶ ADD_DATA UPDATE_INPLACE UPDATE_DISPLAY_NAMES
;
/* ================================================== */
```

Program 5.7-1 Update in place

To program the disable and enable steps, use the OLAPOPERATE procedure options ❷, as shown in the following code:

```
/* ===========================================================*/
PROC OLAPOPERATE;
CONNECT USERID="sasadm@saspw" PW="xxxxxx" HOST="hostname" PORT=5451;
❷ DISABLE CUBE "/Projects/Cubes/Candy_Sales";
 ENABLE CUBE "/Projects/Cubes/Candy_Sales";
RUN;
/* ================================================== */
```

Program 5.7-2 PROC OLAPOPERATE example

 Note that the OLAPOPERATE procedure requires a SAS OLAP Server administrator account. If you run this code without an administrator account, the commands will fail. You can test the planned user account from SAS OLAP Cube Studio and if permissions are not appropriate, the error message **You must have administer permission on the OLAP server to perform this action** appears.

5.8 SAS Administrator Tasks

SAS administrators can use SAS Management Console to establish roles and set security for the OLAP cubes. Cube security can also be defined at the member level to allow different users or groups access to different portions of the same cube.

5.8.1 Granting Access to the OLAP Schema

OLAP schemas organize the cubes for the SAS OLAP Servers. The OLAP schema contains a list of cubes that one or more SAS OLAP Servers can access. Each cube is listed in one and only one OLAP schema.

To grant permission to the OLAP schema, the SAS administrator needs to establish an OLAP Creator group, and then assign the group to the OLAP schema.

- In the User Manager plug-in, establish an OLAP Creator group to set the permissions for users that are creating OLAP data structures. This simplifies administration tasks when users are added or removed.

- To add the OLAP Creator group to the OLAP schema, expand **Authorization Manager > Resource Management > By Location > SASApp-OLAP Schema**. Right-click the schema name and select **Properties**. Add the OLAP Creator group and grant WriteMetadata access. This group has to be assigned only once for each schema.

Figure 5.8-1 Granting access to the OLAP Schema

5.8.2 Member-Level Security

SAS Management Console provides an interface to assign member-level security per user to any member of the OLAP cube.

1. Navigate to the specific OLAP cube by expanding the **Authorization Manager** and navigate from **Resource Management > By Location > SASApp - OLAP Schema**. Then navigate from the cube name to the dimension names and select the properties on a dimension where member-level security is required.

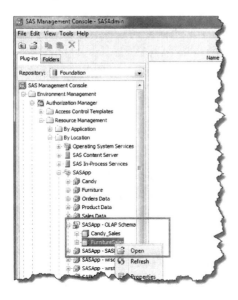

2. On the **Authorization** tab, add the specific user or group that needs member-level security defined. The **Add Authorization** tab ❷ is greyed out. To access the **Add Authorization** tab, click the **Grant Read** check box ❶ to explicitly grant Read access (showing as checked with a white background). Explicit access removes any association with security definitions that are stated above this object in the security model, removing any inheritance between objects above this object and this one.

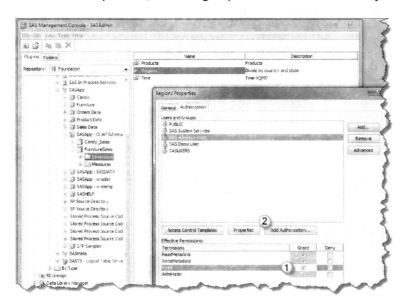

3. After clicking the **Add Authorization** button, you can then specify what members the users can view.

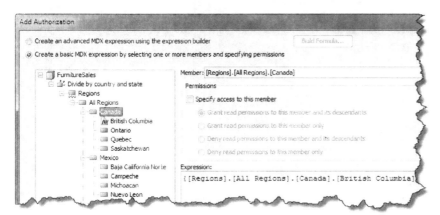

Note: If the cube requires that summarized data at all levels take the security settings into account, than a SECURITY_SUBSET option must be enabled.

At the first window of the cube designer, Cube Designer-General, clear the selection for **Include secured member values in presummarized computations**.

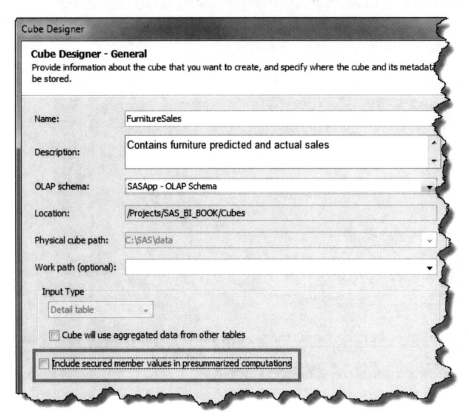

Chapter 6

SAS Information Map Studio

Preparing Data for Report Authors

Chapter 6

SAS Information Map Studio

Preparing Data for Report Authors

Data tables hold valuable information for a business. However, for general users, information such as data table relationships, cryptic variable names, and complex formats can hinder the process of analyzing and generating value from the data. Information maps provide a business layer definition that report authors can understand easily. Key benefits of using information maps include:

- You can customize labels and descriptions

- There is a central location for calculations and filters

- Query and data-storage information remains transparent to report authors

Typically, creating information maps is reserved for the advanced report creators and data administrators who have an understanding of not only the data table structures but also of what report capabilities are needed.

Information maps are used in all of the SAS BI clients. Throughout this book, there are examples of information maps in use in the chapters about SAS Add-In for Microsoft Office, SAS Web Report Studio, and SAS Enterprise Guide.

There are two ways to create information maps: using PROC INFOMAP or SAS Information Map Studio. Many users prefer the simplicity of SAS Information Map Studio. In this chapter, you will use SAS Information Map Studio to create and enhance information maps with filters and prompts. These examples demonstrate how to create a basic information map from relational tables, and then how to enhance the map using filters and prompts. In addition, there are some tips and tricks for solving advanced issues as well as how to use OLAP cubes and SAS Stored Processes as your data source.

6.1 Getting Started

To begin learning about this tool, here is a brief introduction to the tool and what is needed to get started.

6.1.1 Quick Tour

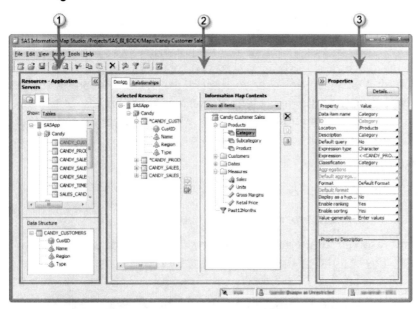

Figure 6.1-1 SAS Information Map Studio display overview

1	Resources pane	Use this area to navigate the data sources that you want to add to the information map.
2	Information Map Creation pane	Use this area to design the information map. You can join data sources and make changes to your information map.
3	Information Map Properties pane	Use this area to quickly make changes to the properties of individual data items.

6.1.2 Prerequisites

Before using SAS Information Map Studio, ensure that the following software is available and all necessary permissions are established:

- SAS Information Map Studio is installed on your desktop.

- You will need permission to access the SAS Metadata Server. Your SAS administrator sets these permissions and provides the name and location of the SAS Metadata Server.

- Data libraries are established and any data sources are registered. The source data used in the examples within this section have already been registered within the SAS Metadata Server.

Note: From SAS Enterprise Guide, use the INFOMAP procedure to generate information maps. Go to support.sas.com to learn more about the INFOMAP procedure.

6.2 Creating Your First Information Map

Information maps provide report authors with a user-friendly way to query data and get quick results. An information map contains data items and filters, which are used to build queries. Setting up an information map is a simple process of selecting and organizing data, customizing new or existing data items, defining filters, and testing the results.

In the following example, an information map about the customers of a candy company is being created. The sales data analyst asked for an information map that she could use to provide reports detailing each customer's sales activity for the past year. The following figure shows the information map final report, as used in SAS Web Report Studio.

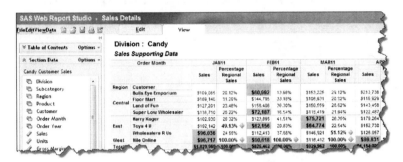

Figure 6.2-1 Information map used in SAS Web Report Studio

6.2.1 Selecting and Organizing the Data Sources

You can select one or more relational tables or a single OLAP cube for the information map to use. To create the Candy Customer Sales information map, use the three SAS data sets shown in the following figure. Each data set contains specific information needed to complete the information map. The first data set, CANDY_SALES_HISTORY, contains a detailed history of the sales; the second data set, CANDY_PRODUCTS, contains the products by category and subcategory; and the third data set, CANDY_CUSTOMERS, list the customers by name. The arrows indicate the data relationships between the data sets. When these tables are joined, the information map is created. For more details about how this report is used to create reports, refer to Chapter 7, "SAS Web Report Studio."

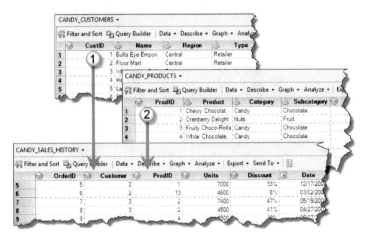

Figure 6.2-2 SAS data set relationships

6.2.1.1 Select the Data Source

To create an information map, do the following:

1. Open SAS Information Map Studio and connect to the server.

2. Use the Resources-Application Servers panel to navigate the libraries to select the tables or cubes that you want to use.

Ensure that any tables or cubes you want to use are registered in the SAS Metadata Server. To learn how to set up libraries and register data, refer to Chapter 2, "SAS Enterprise Guide."

From the **Resources - Application Server**s area, click **Show** and select **Tables** or **Cubes** to filter what the libraries list. For this example, use the SAS data sets from the Candy library.

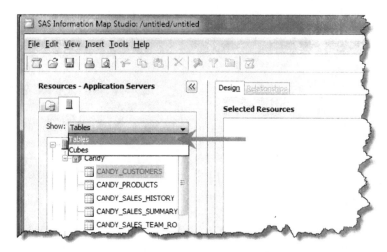

Double-click the data sets to add them to the **Selected Resources** in the **Design** tab. After the data tables are added to the project, the names are listed in **Selected Resources** area in the **Design** tab.

When selecting more than one relational table, define the relationships between the tables for SAS to combine the tables. Click the **Relationships** tab to view the tables. To define the specific join relationships, click and drag from one variable to another to define the specific join relationship.

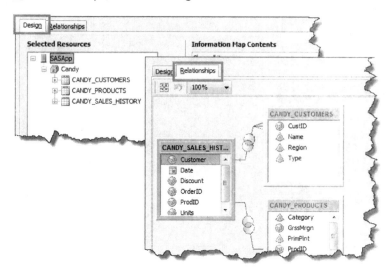

In preceding figure, join CANDY_SALES_HISTORY with CANDY_PRODUCTS by ProdID to get the details of what each customer ordered. CANDY_SALES_HISTORY does not have the specific details about the customers, such as the company name and location. This information is contained in the CANDY_CUSTOMERS table. These tables are joined by the CustID and **Customer** data items. It does not matter if the variables have different names, as long as the values can be used to match. Refer to Section 6.3.2, "Joining Tables," for more details about advanced joins.

3. Add the data items to the **Information Map Contents**. You can also create folders to organize the data items into categories.

   To add a folder, right click the information map name, which is *untitled* because it has not been saved yet. Select **New Folder** from the pop-up menu and type a name for the folder.

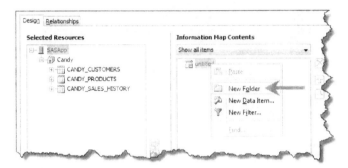

4. On the **Design** tab, select the data items, and then click the single arrow button to move items to the information map. To move data items to a specific folder, select the folder name before moving the data item.

 Use the double arrow button to move everything into the information map at once.

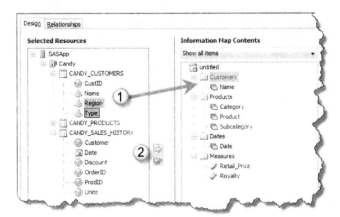

As data items are added to the information map, the icon changes into a category or measure, as shown in the previous figure. The category icons (such as category, product, and name) indicate data items that are used to group measures. The measure icons (such as retail price and royalty) indicate a numerical value. Refer to Chapter 7, "SAS Web Report Studio," for more data-item definitions.

A measure cannot be used to group other data items. Notice in the figure above that the **Date** data item changed from a date icon to a category icon. You can change the names, type, and formats.

For numeric data items that represent qualitative data, change the properties of the item from **Numeric** to **Category**. This ensures that the values are available for grouping in the report.

6.2.1.2 Organizing the Data Items

Data elements can remain in their initial order; however, this is the exact view that a report author sees. To make the information map more user-friendly, reorganize the data items into a logical structure. For instance, in the previous figure, **Subcategory** comes after **Product**. It would make more sense for it to follow **Category**.

1. To reorganize the data elements, use the arrows located to the right of the **Information Map Contents** pane to move data elements around, or click and drag elements within the pane.

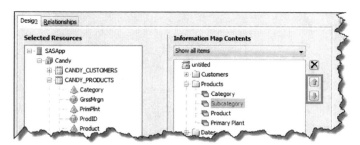

A huge benefit of using information maps is the ability to assign data item names that are familiar to people in the business. In the information map, the **Name** data item actually contains the customer's company name. Because this data item might be confused for salesperson's name or even the customer contact name, rename the data item to **Customer**.

2. To rename the data item, right click the **Name** variable and select **Properties** from the pop-up menu. In the Data Item Properties window, type the new name in the **Data item name** field. The **ID** field contains the original variable name.

The report authors and users see the data item name when they use the information map as a source for reports in SAS Enterprise Guide, SAS Add-In for Microsoft Office, or SAS Web Report Studio. Ensure that the data item name and description use the terminology common to the organization. For instance, some organizations refer to customers as clients, patrons, or even shoppers. In this case, use the common terminology as the name. In the **Description** field, add helpful text about the data item. For instance, add the data source when it originates from an unusual source or list the calculations for **Measures** that might cause confusion for the report builder.

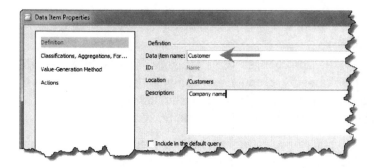

Right click the data item and select **Rename** from the pop-up menu as a shortcut method.

In the example, rename the other data items for clarification and ease; for instance, the **Date** data item became **Order Date** and **PrimPlnt** changed to **Primary Plant**. As you continuing reading this chapter, you might notice other examples of changes made to the data items.

6.2.2 Creating New Data Items

You might want to add new data items to the information map. The data items are based on character, numeric, date, or time columns in a data set.

The Order Date data item is the day the customer placed the order. Because the customers are making many orders in a month, the report author wants the option of grouping the results by month. Using the Date variable, create a new variable called Order Month.

To create the new data item, do the following:

1. There are multiple ways to get to the Data Item interface. Select **New Data Item** from the **Insert** menu, click the **Data Item** icon on the menu bar, or right-click on the folder name and choose the **New Data Item** menu item.

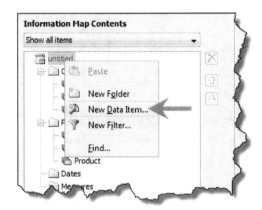

2. The initial screen allows only a new data item name and description. Click the **Edit** button to enter an advanced expression.

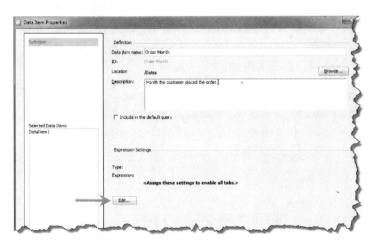

3. In the **Expression Text** field, type the expression based on the data items in the **Data Sources** tab. Use functions from the **Function**s tab, such as INDEX(), MDY(), or CASE statements to create new variables.

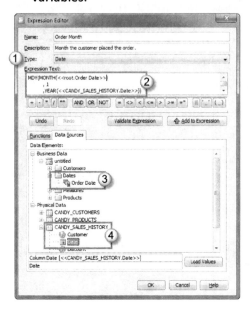

1	Type	Use this field to specify whether the data item is a character, numeric, or datetime value.
2	Expression Text	Use the functions and data items to build the expression.
		In the previous figure, the MDY() function is used to create the new variable. The MDY function sets the day value to the first of the month and uses the month and year values of the Date value so that the data item groups properly in the reports.
		The MONTH and YEAR functions are used to extract the month and year from the order date items.
3	Business Data	Use business data items when the data item from the information map is required. In the last step, Date was renamed to Order Date. Notice that it is referred to as <<root.Order Date>> in the expression. *Root* indicates that it is from business data.
		Use the business data when you created a new data item or otherwise made changes to a data item that could affect the formula.
4	Physical Data	Use physical data items when the data item from the data set is required. Notice that it is called <<CANDY_SALES_HISTORY.Date>> in the expression. This data items uses the data set name to indicate that it is physical data.
		The physical data does not have to be included in the information map to be used in the expression.

 Use the **Validate** button to verify that you typed the expression correctly.

4.   After creating the expression, you might need to set the format for the variable. Click the **Classifications, Aggregations, Formats** choice to make the changes.

In the ❶ **Formats** area, select the ❷ **Format Type** and **Format name** you want, as shown in the following figure.

6.2.3 Defining Filters and Prompts

When using the information map in SAS Web Report Studio, report authors can define filters and prompts when they create a report. However, every time the report authors create a new report based on the information map, they must recreate the same filters and prompts. It makes more sense for common filters and prompts to be available from within the information map.

In this example, the report authors need only the most recent customer order history for reporting. The current information map contains over 5 years of data for all past and present customers. Instead of each report author creating a date and customer filter for each report, add the filter to the information map.

6.2.3.1 Adding a Filter

To add a filter, do the following:

1.   There are multiple ways to get to the Filter interface. Select the **New Filter** item from the **Insert** menu, click the **Filter** icon on the menu bar, or right-click on the map name and choose the **New Filter** menu item.

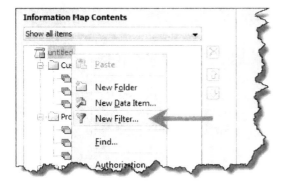

2.   When the New Filter window appears, complete the fields as shown to create a filter that limits the records to the past 12 months.

1	**Filter name** and **Description**	Type the filter name and a short description in these fields. Values typed in this field are what the report authors see.
2	**Data item**	Select the variable that you want to use for the filter.
3	**Condition**	Select how to filter the data item. This selection includes options based on the type of data item filtered, such as **Equal to**, **Not Equal to**, **Is after or equal to**. For dates, the choices include **Year to Date** and **Previous N Periods**. For this example, select the **Is before or equal to** choice.
4	**Value(s)**	Select how the filter receives data. The data can be typed, the report author can be prompted, or the data can be generated from a custom SQL expression. For this example, the filter is one year ago, based on the current date.
5	**Hide from user**	Chose this option when the filter is required to run each time the information map is accessed. See Section 6.3.1, "Setting Up Prefilters," for more information.

6.2.3.2 Adding a Prompt

Prompts are used to allow the end user to choose how to filter the data. In the information map you are creating, the reports might be generated for an individual customer that the end user selects. Create a filter that uses a prompt to select the customer name from a predefined list. Refer to Chapter 4, "The Prompting Framework," for a more complete discussion about prompting.

To create the Customer prompt, use the following instructions.

1. Click the information map name and select **Create a Filter**.

2. To start the filter, type the filter name and select the data item and condition as described in previous section.

3. In the **Value(s)** drop-down list, select **Prompt user for value(s)**. After selecting the method, use the **Values(s)** area to select an existing prompt or create a new one. For this example, select the **New** button to create a new prompt.

4. In the **General** tab, type prompt name and create text to assist with user interaction, as shown in the following figure.

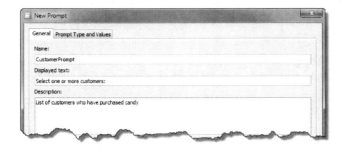

5. Click the **Prompt Type and Values** tab to set up the prompt. For this example, the report author selects a customer name from the list. The customer list is considered dynamic because customers are added and removed, sometimes on a daily basis. It would be time consuming to keep the list accurate. You can set up a prompt to query the table for the customers it contains, thus ensuring that the data is as accurate as possible.

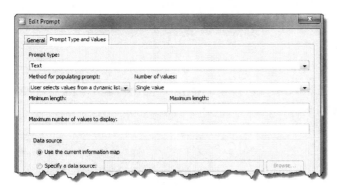

To set up the prompt, do the following:

a. **Prompt type** contains a list of options based on the data element. Generally the default value works.

b. Select how the prompt receives data from the **Method for populating prompt** drop-down list. Choose one of three options: users type a value or can select it from a static or dynamic list.

- When you allow the end user to enter freeform text, use the **Minimum length** and **Maximum length** fields to validate the user-entered text for number of characters. To ensure that the user entered a value, set the minimum length to 1.

- Use a static list when the values do not change often. For instance, for the regions (east, west) or order status (open, shipped). Use a dynamic list when the values change often. In the **Maximum number of values to display** field, specify the number of values initially displayed to the user.

Dynamic lists result from a select, distinct query of the source data table. For large data tables, this query could slow the prompt display.

c. In the **Number of values** drop-down list, specify whether the report author can select one or more, as well as multiple ordered values. Your choice much match the condition you selected for the filter. If you want to allow multiple values, then select the **In a list** item in the **Conditional** field in the New Filter window.

d. In the **Data source** area, select the source for prompt. There are two choices: **Use the current information map** or **Specify a data source**. Use the **Browse** button to find the source you want to use.

e. Click **OK** to create the filter. The prompt is added to the Edit Filter window as shown in the following figure. If you need to make changes to the prompt later, click the **Edit** button.

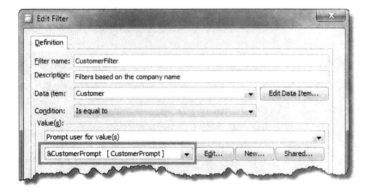

6.2.4 Testing the Information Map

After defining the map, it is a good idea to validate the filters, new data items, and the data joins prior to releasing the information for others to use. Also, you want to save the information map to the designated location. We called this information map Candy Customer Sales.

To test the map, use the following instructions.

1. Select the **Checked Map** icon on the menu bar to open the Test Map interface.

2. In the Test the Information Map window, the **Available items** area lists the data items. The data items can be placed in the **Selected items** area in different combinations so you can review the results. By default, the measures are displayed as aggregated values. By default, 100 rows are shown. To review fewer or more rows, change the value.

3. To complete the test, do the following steps:

   a. Move data items from the **Available items** area to the **Selected items** area using the arrows. Ensure that any new data items and filters are included, such as the CustomerFilter and Order Month.

      Click the **Run Test** button to start the test. Because the CustomerFilter was selected, you are prompted to select a customer, as shown in the following figure.

   b. When the result appears, review any new filters or data items to ensure that they are correct. The Customer filter prompted you to select a customer and Order Month appears in the correct format.

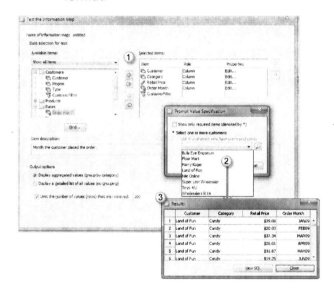

6.3 Enhancing Your Information Map

Using the advanced filtering, joining, and data items techniques available in SAS Information Map Studio, you can shorten the creation and maintenance process.

6.3.1 Setting Up Prefilters

Hidden filters can also be created and then used to prefilter the data source before the user accesses it. For instance, the data set might contain 10 years of data; however, you would like to offer data only for the past 12 months.

One solution is to build a process that creates a new table containing data for the past 12 months alone. Using this approach as a solution, the server quickly fills with small data tables to manage. A better solution is to create an information map with a prefilter that extracts the needed data. This eliminates the need for additional summary tables.

To add a prefilter to the information map, do the following:

1.   Create a filter as shown in Section 6.2.3 "Defining Filters and Prompts." In this example, use the Past12Months filter.

   Select the **Hide from user** option if you do not want the users to see this filter.

2.   Click the information map name and select **Properties**.

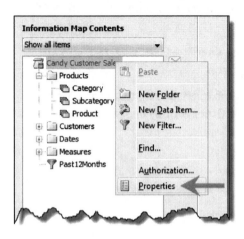

3.   Go to the **General Prefilters** tab. Move the Past12Months filter into the **Selected filters** area, as shown in the following figure.

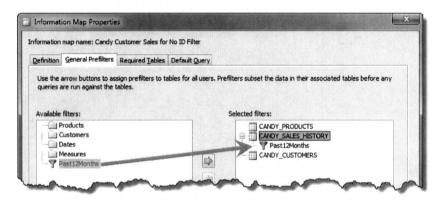

4. Click on the **Required Tables** tab to add the data table connected to the filter. In this example, CANDY_SALES_HISTORY in added to the **Required tables** area.

    Any tables that use a prefilter must be listed as a required table. This ensures that the prefilter is always used, even when no data items from the table are included in the information map contents or the report results.

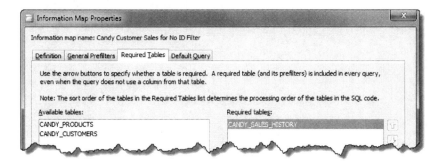

6.3.2 Joining Tables

SAS Information Map Studio can easily be adjusted to assist with improving and adjusting the join behavior.

6.3.2.1 Setting Join Preferences

SAS Information Map Studio can automatically create relationships between tables as you add them into the information map. Open **Tools > Options**. Use the **Initial Creation Rule** area to select the behavior for the joins. There are three choices, as described in the following table and shown in the following figure.

None	Does not create any relationships. This is the default.
From existing relationship definitions	Uses the table metadata relationship based on primary and foreign key designations.
From similarly defined columns	Creates relationships by using table columns that have the same name, type, and length.

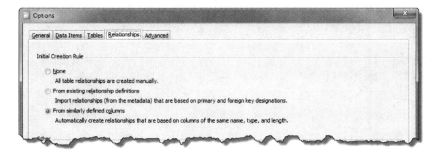

Figure 6.3-1 Setting join preferences

6.3.2.2 Modifying Joins

You can modify the table joins to ensure that the data you want is available.

1.   For each relationship, you also have the ability to change the join properties by double-clicking on the join (Venn diagram icon), as shown in the following figure.

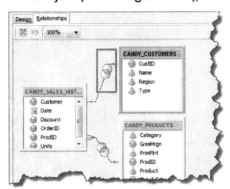

2.   The Relationship Properties window allows you to control how the tables are joined. In this example, CANDY_CUSTOMERS has 1 record per customer, while CANDY_SALES_HISTORY has 1 record per sale. You need to make some changes for the table to join as needed.

   a.   In the **Cardinality** drop-down list ❶, select how the tables are joined. For this example, ensure that the cardinality is **many to one** to indicate there are many orders to a single customer.

   b.   Select the **Outer join** check box ❷ for Table 2. This ensures that all the customers, even those without any orders, are available.

   c.   Use the **Join key**s fields ❸ to indicate how the tables are joined. Use the **Add** button if there is more than one join key.

   The following figure shows an example of creating a many to one relationship between CANDY_SALES_HISTORY and CANDY_CUSTOMERS.

 Use **the Advanced Edit** button if the variables need additional modification for the join.

6.3.3 Adding Custom Formats

Formats are useful when data needs to be sorted in a custom order, grouped differently, or contains values that are not readily understood by the business. Formats can be created either using Base SAS software or using the Create Format task in SAS Enterprise Guide.

 Another example is available in Chapter 7, "SAS Web Report Studio" that demonstrates setting custom sort orders.

In this example, the report authors have requested that a Group data item that categorizes customers, (for instance, convenience or grocery) based on the company name. Instead of creating a new column in the data table for the grouping, apply a user-defined format to an existing variable to create the new variable. When changes need to be made, simply update the format and automatically have the changes applied to your data.

When you create a format, make it available to all of the SAS BI clients by saving it to the Format catalog. The Format catalog is stored in the following default location in the SAS configuration directory:

<configuration directory>\Lev1\SASApp\SASEnvironment\SASFormats\formats.sas7bcat

Use the following steps to create a custom format for the Customer variable.

1. Create the custom format called $CUST_GRPS and save it to the Formats catalog.

    The following program is an example format added to the Format catalog. For more information about the Format procedure refer to the SAS software documentation at support.sas.com.

    ```
    /*=====================================================================*/
    /*Establish connection to the default Format catalog */
    libname FMTS '<configuration directory>\Lev1\SASApp\SASEnvironment\SASFormats';

    /*Create the format and assigned to the FMTS library*/
            PROC FORMAT lib=FMTS;
            value $cust_grps
                    "Bulls Eye Emporium"   = "Convenience"
                    "Harry Koger"          = "Grocery"
                    "Super Low Wholesaler" = "Grocery"
                    "Nile Online"          = "Web"
                    other                  = "Other";
            RUN;
    /*=====================================================================*/
    ```

2. Open the information map where you want to add the variable.

3.   Right-click the data item and select **Copy** from the pop-up menu. In the following figure, the Customer data item is copied.

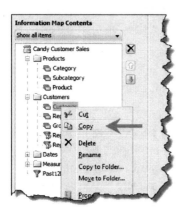

4.   Paste the data item. Then right-click the data item and select **Properties** from the pop-up menu.

5.   In the **Definition** tab, rename the data item. For this example, the data item is renamed to Group.

6.   In the **Classifications** tab, select **User-defined** from the **Format type** drop-down list. All of the user-defined formats appear.

7.   Select the newly created $CUST_GRPS format.

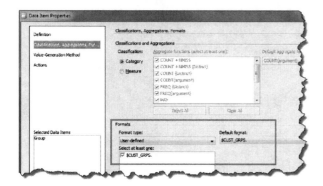

8.   When you test the information map, the new data item called Group has the custom format applied.

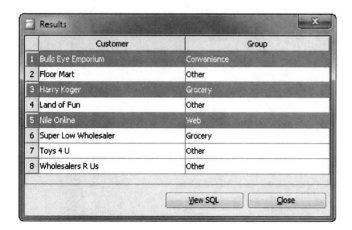

6.4 Tips and Tricks

Prompts and stored processes greatly enhance your flexibility and power when creating information maps. The following tips and tricks provide suggestions for integrating these tools as you develop your information map.

6.4.1 Building Cascading Prompts

For long lists of data, such as all the employees in a large organization, scrolling through the values in a single prompt can be frustrating to users. Creating an initial prompt, such as department, that drives what values appear in the second prompt, and that lists only employees in that department, is what cascading prompts (also called smart lists) are about.

In this example, you are creating a cascading prompt for the region and customer data items. When the report author selects the region, only the customers within that region are available for selection. The basic steps for creating filters and prompts are shown in Section 6.2.3, "Defining Filters and Prompts."

To create the cascading prompt, do the following:

1. Create a new filter called RegionFilter that is based on the region data item.

2. Create a new prompt called RegionPrompt that is based on the region data item filter.

    Note that the method for populating this first prompt can be any of the available options; however, the remaining (cascading) prompts must be set at **User selects values from a dynamic list**. In the following figure, the chosen population method is set to use a static list for the first prompt because the three regions (Central, East, and West) rarely change.

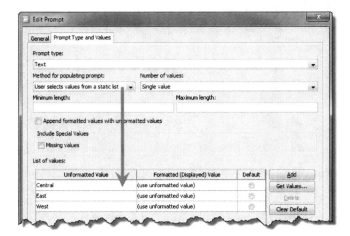

3. Create the next filter and select **Prompt user for value(s)** and then **New to create the second prompt**.

    Give the filter a name that helps you and users identify its purpose later.

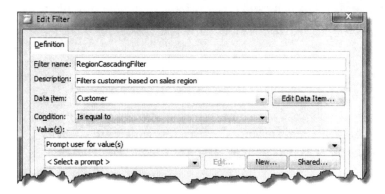

4. As you create the prompt for the customer, ensure that the RegionCascadingPrompt prompt uses the method **User selects values from a dynamic list** and the data item is the data item you want available. In this example, **Customer** is the data item.

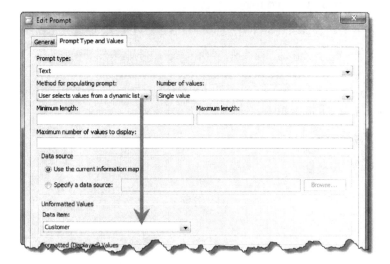

5. After creating the prompt, the Edit Filter window appears. Use this window to set up the combination filter.

    a. Select the **Combinations** button to expand the **Filter combinations** area if it is not already expanded.

b. Select the **Add** button to move the new filter to the **Filter combinations** area. Notice the filter combination Customer = (&RegionCascadingPrompt) indicates how the Customer data item is filtered.

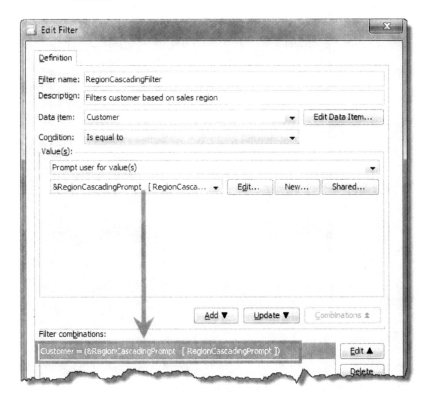

c. The other part of the combination filter is the Region prompt. Modify the data item selection to Region and change the Value(s) area to point to the &RegionPrompt prompt. Click the **Add** button.

d. Within the Filter Combinations area, select the **Region= row** and use the up arrow button beneath the field to move the Region to the top. **Region** should appear first, as shown in the following figure.

e. Check the **Establish dependencies between prompts** check box. This ensures that the filters are used together.

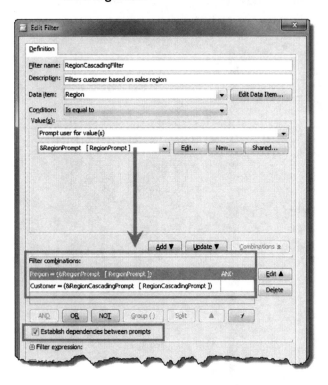

6. Test the cascading filter using the **Test Results** menu item. Refer to Section 6.2.4, "Testing the Information Map," for detailed steps.

When testing the new prompts in SAS Information Map Studio, the second prompt values populate with options after the first prompt value is selected. In the following figure, four data items are selected: Order Year, Region, Customer, and Sales. The test feature allows only one prompted filter to be tested at a time. Make sure you select the cascading filter called RegionCascadingFilter.

As you can see, the Prompt Value Specification window shows the new cascading filter in action. For the East region, there are only three customers available. The Results window shows the sales summarized for the Toys4U customer in the East region.

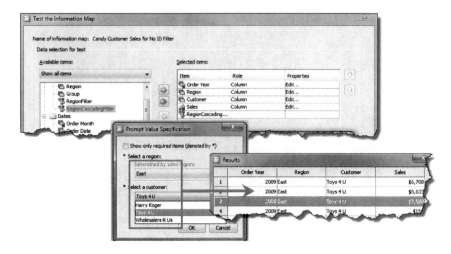

6.4.2 Using Shared Prompts

Prompts can be shared using SAS Management Console so they are reusable between multiple information maps. There are several requirements for your map to use a shared prompt successfully.

- The prompt must require a value. Select the **Requires a non-blank value** check box when creating the shared prompt.

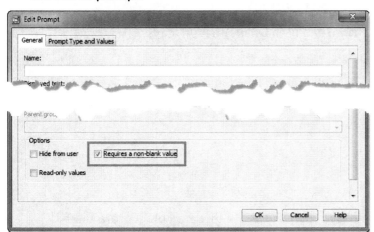

Figure 6.4-1 Shared prompts options

Once this requirement has been met, the shared prompt is available from the **Tools > Manage Prompts** menu. Use the **Add Shared** button to make the prompt available to your map.

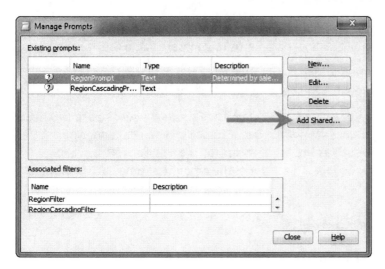

Figure 6.4-2 Adding shared prompts

6.4.2.1 Limitations of Accessing Shared Prompts

There are several situations where shared prompts cannot be directly used in the manner described in the previous section. These constraints can be worked around by setting up prompts in stored processes and adding the stored process to the information map. Refer to Section 6.4.5, "Integrating SAS Stored Processes" for more information.

- Information maps cannot directly use shared prompt groups.

- The following prompt types cannot be shared and used by information maps.

  - Color

  - Data Source

  - Data Source Item

  - File or Directory

  - Data library

  - Text

    Note: SAS Information Map Studio does support single-line text prompts that are shared.

- Shared prompts cannot include dependencies. This means that one prompt cannot have values populated based on another prompt selection.

6.4.3 Allowing Value-Generation for Prompts

Within SAS Web Report Studio, report writers can create filters and prompts to use within their specific reports. To improve the information map usability, allow data items to generate value lists for the filters.

To change the **Value-Generation Method** option for each data item, click the data item and select **Properties**. In the Value-Generation Method pane, select one of the three methods.

In the following example, the Customer data item allows a dynamic list. This makes sense because the customer list changes over time as new customers are added. Use a static list when data does not change (or at least it does not change often), such as the list of states for the United States, or the environment names (Development, Test, Production). This reduces the amount of time each prompt takes to appear in the report.

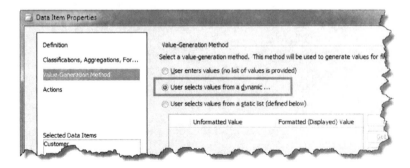

Figure 6.4-3 Allow Value-Generation prompts

Review each data item to determine the ones you would expect report author to create in filters; for example, region, company, and product are likely categories for filters. Report authors are less likely to use non-categorical data items, such as street address and order date, for filters; those data items would not require this option.

6.4.4 Using CASE Statements

The final code that is run against the data table is SQL based, which means that you can also create custom data elements that use CASE statements. This is similar to IF/THEN/ELSE logic used by programmers in the SAS DATA step.

CASE statements are useful when you want to generate totals based on the values of the data items. For the following example, a data item is created that has a value of 1 or 0 depending on the value of the Region column. The resulting report shows the total number of records within that particular region.

To create an expression that uses a CASE statement, do the following:

1. Create a new data element.

2. Add a name for the data item and select the **Edit** button within the **Definition** area.

3. In the expression editor, type the CASE statement in the **Expression Text** area, as shown in the following figure.

    Use the **Functions** tab to locate the CASE function example and click the **Add to Expression** button to place the sample text in the **Expression Text** field.

 Navigate to the data item in the Data Sources tab and select the Load Values button to display the distinct data item values.

The following CASE statements use two conditions to create a counter.

Plain English	CASE Statement

If status equal Open and units are less than or equal to 10,000, then count as 1, else count as zero.	`CASE <<root.status>>`
	`when 'Open' AND <<root.Units>> <= 10000 THEN 1`
	`ELSE 0 END`

If status equals Closed and Order Date is within the current month, then count as 1. If the ELSE statement is missing, the value is set to missing.	`CASE <<root.status>>`
	`when 'Closed'`
	` AND <<root.order_date>>=INTNX(month, today(), 0) THEN 1`
	`END`

6.4.5 Integrating SAS Stored Processes

SAS Stored Processes can enhance information map capabilities. SAS Stored Processes are SAS programs that can be run on demand. Stored processes can perform a variety of tasks, such as generating data sets, building queries, and so on. Almost anything you can do in a SAS program, you can do with stored process. Refer to Chapter 3, "SAS Stored Processes," for more information about creating stored processes.

6.4.5.1 Including Stored Processes

To use a stored process with your information map, do the following:

1. Create the stored process and set it to run in the logical workplace server.

2. From SAS Information Map Studio, start a new information map and add at least one data source to the map.

3. From the **Resources - Application Server**s area, click **Show** and select **Stored processes**.

4. Double-click the stored process you want to add. The stored process is displayed in the **Selected Resources** area. In the following figure, the IMAP Group Prompt stored process was added. In this example, the stored process is used to refresh the sales data tables.

6.4.5.2 Using Stored Processes with Information Maps

Stored processes can perform a variety of tasks to make your information maps more powerful and automated. Examples of potential uses within an information map are:

- Connecting to a relational database management system (RDBMS)

    The default access method to RDBMS is through the implicit Access Engine and a SAS library reference. However, when using certain SAS functions, the Access Engine is not equipped to translate into RDBMS native functions. This can cause the SAS server to retrieve all the records in the queried RDBMS data table and process everything in SASWORK. For small data tables or large server environments, this is usually not an issue. However, when working with extremely large tables or constrained SAS servers, it is necessary to build explicit RDBMS queries within SAS Stored Processes and use them within the information map.

- Querying an OLAP cube to allow for prompting (prior to SAS 9.2)

    In prior versions of SAS Business Intelligence, you were unable to use prompts with OLAP cubes. However, using a SAS Stored Process, you can then provide the results within the information map for report authors to surface data in SAS Web Report Studio.

- Dynamically determining which tables are joined together by using prompts and a query within a stored process. This increases the available prompting options.

    An example is shared prompt groups, which cannot be directly accessed in information maps. Multiline text, color, data source, and other prompt types are not available in information maps. To leverage these prompts, you must use a SAS Stored Process that has these associated prompts. Refer to Chapter 4, "The Prompting Framework," for more information.

6.4.6 Sourcing Maps with OLAP Cubes

As mentioned in Section 6.2.1, "Selecting and Organizing the Data Sources," you can use an OLAP cube as the source data for an information map. Adding an OLAP cube into the information map requires that you first select the OLAP type from the drop-down box in the Resources pane, as shown below.

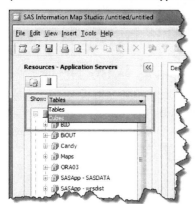

Figure 6.4-4 OLAP cube as source

Prior to using any OLAP cube as a source for information maps and subsequent web reports, it is important to understand what this source format provides.

Pros	Cons
Drill into and expand tables and graphs	Unable to use list data report object
Drill into detail data from the report with no special coding	Cannot link to OLAP source report from another web report
Support ragged and unbalanced hierarchies	Information maps can contain only a single cube
Switch dimensions and measures with the Data Selection window	Able to consume the shared prompt only with defined type as OLAP Member
Synchronize report components to display a common drill state or have them remain independent	Identity based filters are assigned at the cube and cannot be created within the information map

6.4.7 Creating an Identity-Based Filter

When some records in the table should or should not be viewable to a select group of users, you might create different information maps to meet each group's needs. However, when this number exceeds more than a couple, it makes sense to reduce duplication and create a single information map that uses information stored in the metadata server for each user to filter the data table to display a subset of records.

SAS Information Map Studio has six identity properties to assist with row-based filtering. These filters allow you to be as general or specific as desired. For example, use an Identity Group for all members of a group, like Sales Department, or assign security to a specific property, such as User ID, like Angela Hall. The following table describes the identity properties. Refer to Section 6.5.4, "Identity-Based Filter Properties," for information on how these properties align to the User and Group information in SAS Management Console.

Properties Available	Description	Example
SAS.PersonName	User name as defined in SAS Management Console	Angela Hall
SAS.IdentityGroupName	Group name as defined in SAS Management Console	Note – This property is used only when a group login ID is accessing the information map.
SAS.IdentityName	User or group name as defined in SAS Management Console	Angela Hall
SAS.Userid	Authenticated user ID	Anhall
SAS.IdentityGroups	List of groups and roles in which the metadata user is a direct or indirect member	Sales Team - East

Properties Available	Description	Example
SAS.ExternalIdentity	A site-specific external identifier	AAL_Group, PSD_Community **Note:**Only the first value for external identities is returned for use in the filter.

Table 6.4-1 Identity-based filters

In the following example, you are setting up an Identity Group filter for the Region data item. Each sales team should be able to see the customers and sales data only for their assigned region.

Note: This example uses relationship data tables and does not work for OLAP cube data.

Use the following steps to create a new identify-based filter:

1. Either add a new data item to your relational data table or create a new table with the user identities. This step requires modifications to the data table. In the following figure, a Role data item was added. Role is based on the values in Region. Each sales team is added to the region they represent.

   Note: These groups must also be set up in SAS Management Console. Refer to Section 6.5.4, "Identity-Based Filter Properties," for more information about the groups.

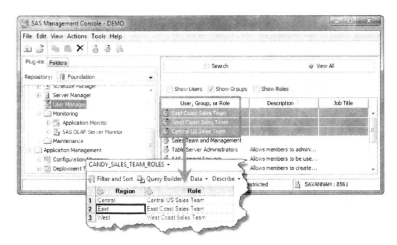

2. Open the information map and add the new data table. In this example, the data table is called CANDY_SALES_TEAM_ROLES. Go to the **Relationships** tab and join the new data table to the appropriate data table. In this example, Region is the common data item.

   It is not necessary to add any of the data items in the CANDY_SALES_TEAM_ROLES to the information map.

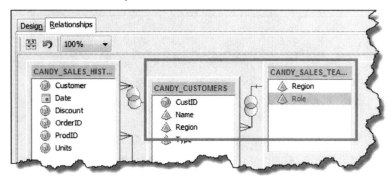

3. Create a new filter in your information map. In the Edit Filter window, do the following:

   a. In the **Filter name** field, type the name, such as IdentityFilter.

   b. For the data item, select **Edit Data Item** to open the Edit Expression window.

   c. In the **Type** drop-down list, select **Character**. In the **Data Sources** tab, expand **Physical Data** and add Role to the **Expression Text** field as shown in the following figure.

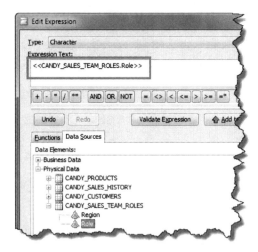

   d. In the **Value(s)** field, select **Derive identity values (for row-level permissions)**.

   e. Click on the identity property you want. Because this filter is at the group level, the SAS.IdentityGroups property is selected.

f. Click the **Hide from user** check box so the report author cannot see or make changes to the filter. This also completely secures the information map.

4. Go to the **Properties** tab for the information map and add the newly created Identity filter as a prefilter, as shown in the following figure. Also, add the CANDY_SALES_TEAM_ROLES table to the **Required Tables**. Refer to Section 6.3.1, "Setting Up Prefilters," for more information about creating a prefilter.

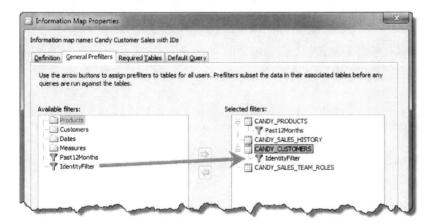

5.   Run a test query.

The following figure shows the information map being used in SAS Web Report Studio with different permissions. SAS User is a member of the East and West sales teams, and SAS Demo User is a member of only the Central sales team.

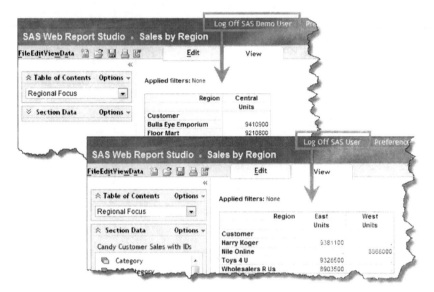

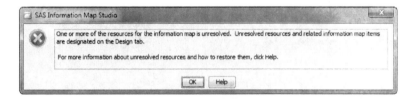

6.4.8 Repairing Broken Information Maps

Use the Resource Replacement feature to update an information map when changes occur. SAS Information Map Studio alerts you immediately about issues. Upon opening the broken information map, the following error message appears, indicating there are unresolved resources.

Figure 6.4-5 Unresolved resources error message

If the data source itself changes (such as the physical folder is deleted), the error occurs when you test the map or a user tries to use the information map.

When you view the information map, red Xs appear with the data resources and data tables, indicating where the issues exist. For instance, in the following figure, three source tables, Customers, OrderDetails, and ProductLines, have unresolved resources. The Orders and Payments tables are correct. Also notice that individual data items in the **Information Map Contents** area are affected.

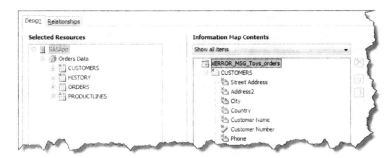

Figure 6.4-6 Information map with issues

To replace the resources with valid ones, do the following.

1.   Select **Tools > Resource Replacement**.

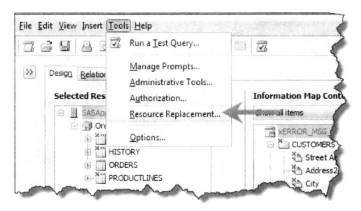

2. When the Resource Replacement window appears, red exclamation points guide you to the areas where the issue exists. In the following figure, the **Tables** and **Columns** ❶ need attention.

Select each red exclamation point area to display the broken resource. Indicate a replacement library or table and then use the drop-down list to select the new resource. In the following figure, the Customers data source❷ was causing the issue because the data source was renamed to Customer.

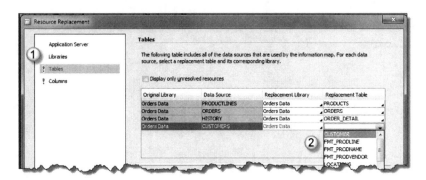

After selecting the replacement table, the red exclamation point is removed automatically because the columns are the same in this example.

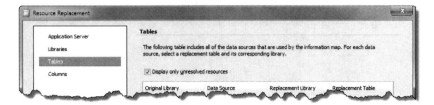

This technique works extremely well when metadata changes, but you can also use it when you want to modify the information map source tables. Using the Replacement Resource feature ensures that the resulting report, such as in SAS Add-In for Microsoft Office or SAS Web Report Studio, continues to work without recreating the entire report.

6.5 SAS Administrator Tasks

Using SAS Management Console, the SAS administrator can set responsibilities, establish folder structures, and define user properties.

6.5.1 Roles and Responsibilities

There are at least two roles for information maps:

*   Users of information maps require ReadMetadata access to the folders where the information maps are stored.

- Developers of information maps must have WriteMemberMetadata and WriteMetadata access to the location within **SAS Folders** to save, share, and edit. In the following figure, a group called IMAP Developers was established and given access to the **InfoMaps** folder.

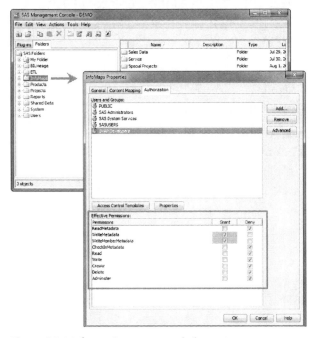

Figure 6.5-1 Information map permissions

6.5.2 Organizing Metadata Folder Structure

You should provide separate metadata folders for the data, information maps, stored processes, and prompts. These separate folders can exist within parent folders for each organization, project, or environment.

In the Candy Company, the organization has structured the folder as represented in the following figure. Information maps are stored separately from reports or stored processes. This allows security settings for different user communities.

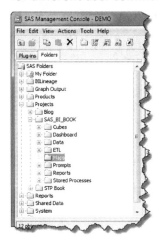

Figure 6.5-2 Metadata folder structure for information maps

6.5.3 Sharing Optional Prompts

For shared prompts to be available in information maps, they are required to have a value (using the **Requires a non-blank value** option). But if the prompts need to be optional for the user community, some users would rebuild the prompt for each information map that uses it. You can share prompts and include a built-in option to select all values within the prompt.

To make a shared prompt optional, do the following:

1. Within SAS Management Console, create a Prompt Manager stored process and create a new prompt. Refer to Chapter 4, "The Prompting Framework," for more details.

2. Move to the **Prompt Type and Value**s tab. In the **Include Special Values** area, select the **All possible values** check box. The following figure shows the check box when dynamic or static is selected.

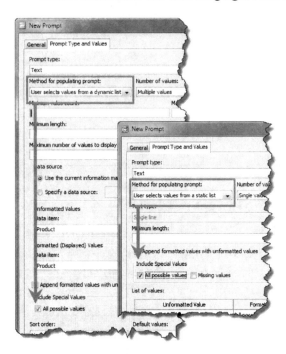

3. Now after sharing the prompt, add it to the information map by clicking **Tools > Manage Prompts**.

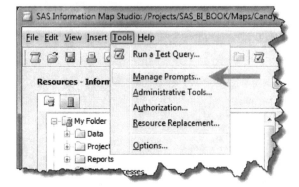

4. Then select the **Add Shared** button, navigate to the shared prompt, and add it to the information map.

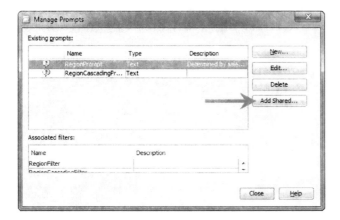

The **All possible values** option is not compatible with cascading filters because cascading filters require a value to subset the subsequent prompts.

6.5.4 Identity-Based Filter Properties

Creating identity-based filters requires an understanding of where the identity properties are defined. They are defined solely in the User Manager plug-in for SAS Management Console. From the User Manager, right-click the user and select **Properties** from the pop-up menu.

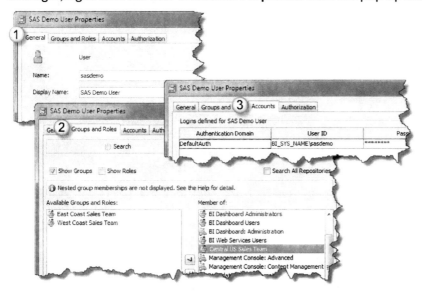

Figure 6.5-3 SAS Demo User Properties

1	General	Shows the **Name** (SAS.PersonName) and **Display Name** (SAS.IdentityName)
2	Group and Roles	Lists all of the groups and roles (SAS.IdentityGroups)
3	Accounts	Defines the **User ID** (SAS.UserId)

Chapter 7

SAS Web Report Studio

Quick and Easy Reports for Users

Chapter 7

SAS Web Report Studio

Quick and Easy Reports for Users

Allowing a large group of users to access data and providing an easy way to do so is paramount to a successful BI deployment. SAS Web Report Studio fulfills the need with an interface and functionality that takes someone just a couple of minutes to understand and only a few more to become proficient in creating their own new reports.

Essentially, SAS Web Report Studio is a reporting tool; however, your organization will find that this is a data access tool with broad appeal. Querying data is usually relegated to programmers or analysts because this group has advanced experience with programming tools such as Base SAS software and SQL. While SAS Web Report Studio does not have the same level of flexibility and functionality as other SAS tools, it can act as a data access tool, allowing anyone to quickly query data and analyze results. The allure expands to anyone interested in making a decision based on data.

Broad appeal can increase the concern for security and training. It can take only a few hours to receive initial training on this product. There is no programming required because everything within the application is in a point-and-click interface. On the security side, administrators can control which data is available and to whom, whether users can view reports only, or they can also create reports and where reports can be stored.

7.1 Getting Started

To begin learning about this tool, here is a quick overview of what it takes to get started with SAS Web Report Studio.

7.1.1 Quick Tour

SAS Web Report Studio has two main modes that you can access from tabs on the main window, as shown in the following figure.

- Edit mode allows you to create or edit reports. This tab is available only if you have authorization.

- View mode displays the finished report.

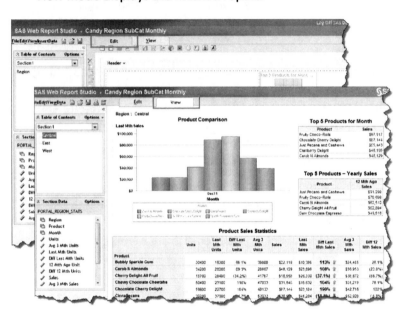

Figure 7.1-1 Overview of the main window

7.1.2 Prerequisites

SAS Web Report Studio is a Web-based application. Your SAS administration can give you the specific server address. The URL address format is as follows. This address is case sensitive.

```
http://server name:port number/SASWebReportStudio
```

If you are creating or viewing reports, you must have a user ID and password identified in the SAS system. The SAS administrator sets permissions and determines which data libraries and sources are available. Section 7.6, "SAS Administrator Tasks," contains more information about SAS Web Report Studio roles and responsibilities.

7.2 Understanding Report Data

To create a report, you need your data in a structure that SAS Web Report Studio can display. The following section defines data sources, icons, and data types. Refer to Chapter 5, "SAS OLAP Cube Studio," and Chapter 6, "SAS Information Map Studio," for more information about creating these sources.

7.2.1 Understanding Data Sources

There are three data sources available to SAS Web Report Studio: information maps, OLAP cubes, and relational data tables. By default, only information maps are initially available; however, relational data tables and OLAP cubes are accessible directly if enabled by the SAS administrator.

These data sources are identified as relational (two-dimensional) or multidimensional.

Icon	Description
Information Map	This is a generic information map icon. Refer to Chapter 6, "SAS Information Map Studio," for more information on creating information maps.
OLAP Cube	This icon identifies an OLAP cube, which is a set of data that is organized and structured in a hierarchical, multidimensional arrangement. Refer to Chapter 5, "SAS OLAP Cube Studio," for more information about creating OLAP cubes.

7.2.2 Understanding Standard Data Items

Each data source includes one or more standard data items. You decide which data items to use to define a query for a report section. You can use all the data items in the data source or a subset of data items. Each standard data item is classified as a category, a hierarchy, or a measure.

Data Item	Definition
Category	Indicates a Category data item with distinct values that are used to group and aggregate measures. There are four types of categories: alphanumeric, date, timestamp, and time. Examples of alphanumeric categories include data items such as Product ID, Country, Employee Number, and Employee Name. Date, timestamp, and time category examples are Order Year, Date of Sale, and Delivery Time.
Measure	Indicates a Measure data item that contains values used to complete calculations, such as sales revenue, days between dates, and salary. Each measure has a default aggregation method. Examples of measures include counts, average, or sum.
Hierarchy	Indicates a Hierarchy data item that is an arrangement of the Category levels in a dimension from general to specific. For example, a common dimension is Time; a dimension hierarchy named YrQtrMth would expand the date from year to month.

7.3 Creating Your First Web Report

SAS Web Report Studio provides many different ways to build a report. As you start, you can use the templates to create reports. However, it does not take long to create your own reports from a data source. The following table summarizes the different techniques for creating a new report.

Even if you use a quick-start method, you can make edits to the report until it meets your needs.

Method	Description
Report Wizard	For brand new users, the Report wizard is the fastest way to generate a report. This feature guides you through the steps needed to use the application.
Stored process as the template	Stored processes can be opened directly from the Open Report interface or included as a part of an existing report. For more information about stored processes, refer to Chapter 3, "SAS Stored Processes."
Template with your data	Standard layout designs with the same elements are saved to templates and used to create reports quickly. A common design is a report with a custom header and footer and one table.
	If another report element is needed, an alternative template must be selected, a new template created, or the report author must use a different method to create the reports.
Create from a data source	When you save a report based on a data source, you create a new report simply by selecting an information map, a table, or a cube. Because the resulting report is based on a data source, you have full access to all the report objects.
Use an existing report	Any existing reports can be opened and saved with a different name or within a different location. This new report can be edited to use a different data source or different report elements.

7.3.1 Using the Report Wizard with an OLAP Cube

The Report Wizard guides you through five steps to create a one-section report quickly. After using the wizard, you can modify the report for any customizations you want. For this example, you work for a furniture company and need to create a regional report that evaluates the predicted and actual sales. This example uses an OLAP cube that allows the end user to see the regional performance and then expand the report to see the performance for each year.

Note: Information on how to build the FurnitureSales cube is in Chapter 5, "SAS OLAP Cube Studio."

To use the Report Wizard to create a report, complete these steps:

1. Select **File> New> New Using Report Wizard**.
2. Use the **Select Data Source** button to locate a data source to create your report.

   Select the data items that you want to use in the report. In the **Available data items** box, select one or more data items and click the arrow icon to move them to the **Selected data items** list box.

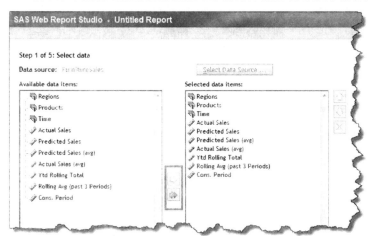

Click the **Next** button to go to the next wizard page.

   At this point, instead of clicking **Next**, you can click **Finish** to display the **Edit** tab. Defaults are used for any remaining unspecified required content. For example, if you are using a relational data source, a list table is automatically included.

3. Step 2 of the wizard allows you to change the formatting on some data items. Typically, the data formats are acceptable and you do not need to make any changes.

 Click **Preview data** to see the data.

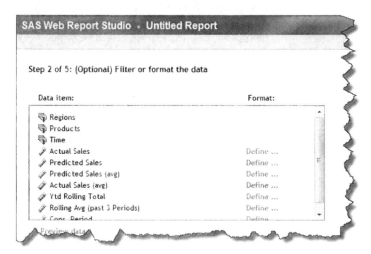

4. Click the **Next** button to go to the next wizard page.
5. Step 3 of the wizard allows you to create group breaks for your report.

A *group break* is a way to divide report sections by distinct category or hierarchy level values when you are using a relational or multidimensional data source.

Complete these steps:

a. From the first **Break by** drop-down list, select a category or hierarchy to specify the first group break.

    If you want the report to display a new page for each value in the first break, select **New page for each value**.

b. Specify up to two more breaks.

    If you do not want labels to appear with each value, then clear the **Label each value** option.

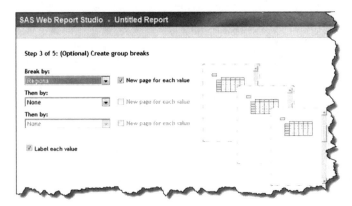

6. Click the **Next** button to go to the next wizard page.
7. Step 4 of the wizard allows you to select how you want the data displayed in the report.

    Add at least one view element to display the results of the query, either a table or a graph (a bar chart, a line graph, or a pie chart). The data items that you selected in Step 1 of the wizard are shown as selected to display. Either accept the default assignments or clear the **Show** check box to hide the data items that you do not want to display.

    - If you select the **Table** option, select the type of table. If your data source is multidimensional, then your only option is **Crosstab**.

    - If you select the **Graph** option, select the type of graph.

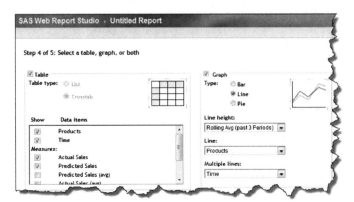

Click the **Next** button to go to the next wizard page.

8. (Optional) To add a header or a footer, complete these steps for each feature that you want to include in your report:

From the **Banner** drop-down list, select an image to use for the banner. Available images are included into the server by your SAS administrator. If there are no images available, then your only selection choice is **None**. Refer to Section 7.6.2, "Adding Images and Logos," for more information about adding artwork to SAS Web Report Studio.

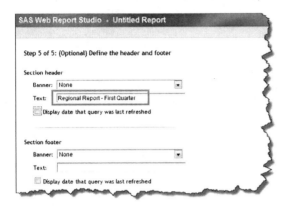

a. Type the text that you want to include in the header or footer.

 You cannot use the following punctuation characters in headers and footers: < > & #

b. To include the date that the section query was last run, select **Display date that query was last refreshed**.

9. Click **Finish** to display the report in the **Edit** tab. The following figure shows how the report looks in the **Edit** and **View** tabs.

Click the **View** tab to see the results. In the following report, each product has an individual table and graph. This report is showing the regional sales for the Boots product line for all of Canada.

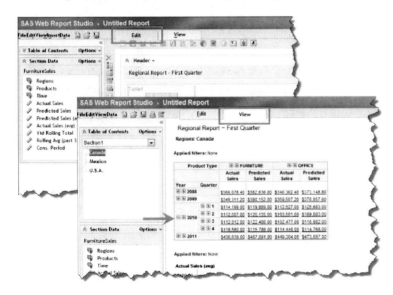

7.3.2 Using a Template with an Information Map

Another way to create a report is to start with a basic template that you can later customize.

This example uses data from a candy sales company by using an information map. This report shows the sales by division, product lines, regions, and customers for the past year. This report is updated every month with the new sales data. The Sales department has requested this report so they can determine how accurate their sales predictions are for the various products and regions.

Note: Create this information map by following the example in Chapter 6, "SAS Information Map Studio."

To create the report, do the following:

1.   Select **File > New Using Template** to start creating the report.

     There are 12 templates available to help you get started. Your organization might also have templates made by others that you can select from **Shared Templates**. You can create templates, which are available from **My templates**.

2.   From the **General templates** tab, select **Two graphs over table**, as shown in the following figure.

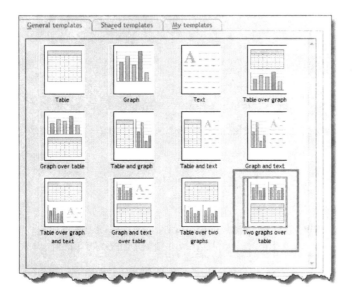

3.   This template has two bar charts above a general report. Above our new template objects, several icons show sample report layouts. For instance, you can choose to have a pie chart, scatter diagram, an image, or even just text in the report.

For this report, you need a line graph and a crosstab to replace some existing elements. Right-click the bar chart called Graph2 and select **Remove chart**. Then drag the line graph icon into the empty area. Replace Table1 with the crosstab icon using the same method. After making the changes, the template now looks like the following figure.

 When the yellow triangle icons appear, it is not an error but a warning to remind you that more information is required for the object to display successfully.

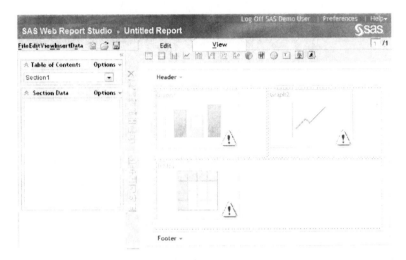

4. When you replaced the table with the crosstab object, the bottom row became two cells. You can control the layout and alignment of your report using the icons to the left of the report area. Select the ❶ two columns in the bottom row and select ❷ the **Merge cells** icon to create a single row.

 Use the other icons to add or remove cells or align an object within the cell.

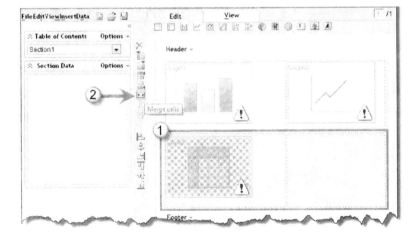

5. To add data to the report, click **Data > Select Data** to display the data source window. This example uses the Candy Customer Sales information map.

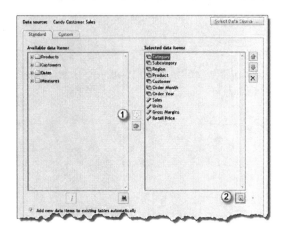

1	You can use the arrow in the middle of the window to move the data to the **Selected data items** column. For this example, the data source Candy Customer Sales, which is an information map, is selected. Move all desired data items to the **Selected data items** column. The icon next to each data item name indicates the data type.
2	The data items might not match what you want for your report. For instance, the data table uses the label Category but for this report, you need the name to reflect how the business is organized. In this case, Category should be Division. To change the data item name, click the data item and select the icon in the lower right corner. In the window that appears, type the new name. For this example, change the name to Division.

6. After adding the data source, SAS Web Report Studio assigns it to the objects. Click **View** to see the result, which is similar to the following figure.

  Save the report so if you do not like the changes you make, you can easily return to this view.

At this point, this report is not meaningful and it would be difficult to do any useful analysis with this information because it is hard to see any trends or understand the sales figures. In the following topics, you will learn how to make modifications to this report to further enhance the data and improve navigation.

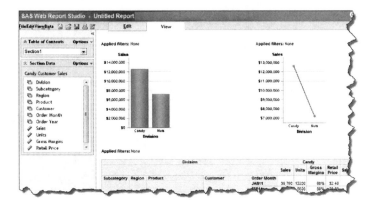

7.3.3 Grouping the Data

The candy company has two categories of products, candy and nuts, and a division manager is assigned to each. You are creating this report for each divisional manager. It shows the sales trends by month and determines which products are selling the best.

Using the Group Break feature, you can create one report and then display the data for each division. If you need to make changes to the report, it is easy. Instead of creating separate reports, you can create one report, with a page for each category.

1. To group the report data, select **Options > Group Breaks**.

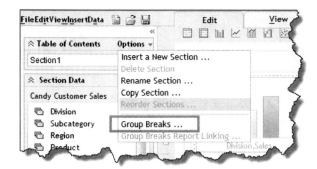

2. For this example, select **Division** and the **New Page for each value** check box. The **Sort** radio buttons allow you to determine how you want the Division names sorted.

   You can also choose the formatting for this title in the window. Because this is heading text, select a 12 point or above size. You can also control the font type and color.

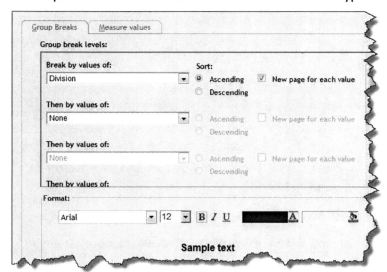

   After you create the data grouping, the window changes. **Division** is now **Section 1** in the Table of Contents pane. The **Section Data** area shows all categories and measures.

3. It is a good practice to change the name of the section from Section 1 to something more meaningful, such as Division Reports.

To rename the section, select **Options > Rename Section**. In the Rename window, type the desired name, such as Division Reports. When you return to the window, the name has changed from Section 1 to Division Reports, as shown in the following figure.

7.3.4 Working with Bar Charts and Line Graphs

When adding data, SAS Web Report Studio automatically assigns the data items to existing elements in the report. The default assignments might not always suit your needs. You can change how the data appears in order to customize your report.

1.   To modify the chart object, right-click **Graph1** and select **Assign Data** from the pop-up menu.

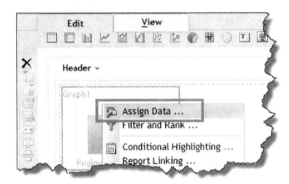

 If you select **Data source details** on any object within the report, you can view information on the type of data, its description, and any applied filters.

2.   For this bar chart, you want to see sales for each product.

Each data item uses a specific value. For instance, **Bar Height** expects to have a measure assigned to it. The ruler icon appears.

If the data item expects to have the categorized data assigned, the 3-box icon appears.

If an item is not available for assignment, it appears as gray font. In the figure, the **Move to Bar Height** appears in gray.

3. Assign the Sales measures to the Bar Height data item. Place these items in the order you want them to appear in the chart.

4. Assign Product to the Bars data item.

 You can use the Move Items menu to help assign the data.

5. For the line chart, you want to show the sales broken out by subcategory per month. Right-click the line graph object and select **Assign Data** from the pop-up window.

   As with the Bar Chart object, each data item uses a specific value. Line Height expects to have a measure assigned and Line expects to have a category assigned.

6. Assign the Order Month to the Line data item. This is the horizontal value on the chart. Assign Subcategory to the Multiple Lines data item so there is a line for each subcategory.

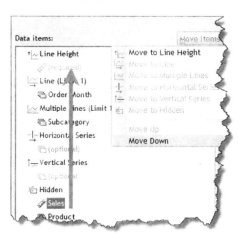

If you peek at the report, the bar chart is crowded. It might be more interesting to just show the top five products sales for the year.

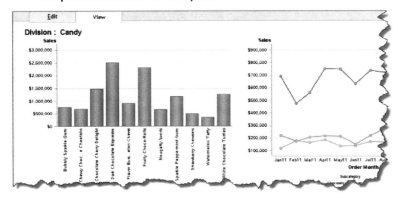

7. To rank the data in the chart, use the following steps:

 Rank is available only when the Synchronize Objects feature is disabled.

a. Right-click the bar chart and select **Filter and Rank** from the menu.

b. From the **Measure Filter or Rank** tab, type 5 in the **Top** field.

c. Select **Sales** in the **Measure** field. You could also select another measure, such as **Units**.

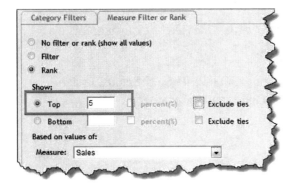

8. The Properties window allows you to add titles, control the object size and customize other object elements.

a. Select the **Edit** tab to set the title.

b. Right-click the graph and then select **Properties** from the pop-up menu.

c. In the General pane, type the chart name in the **Text** field. You can change the look and placement of the title if you want.

d. In the **Graph size** area, click **Fixed size** and select **Custom** from the drop-down menu. Type the width and height values you want.

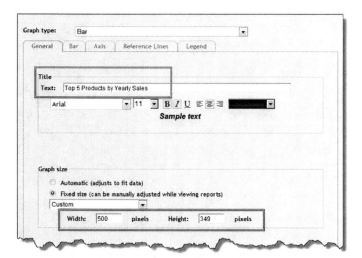

9. After setting the titles and size for the other item, click the **View** tab to preview the report.

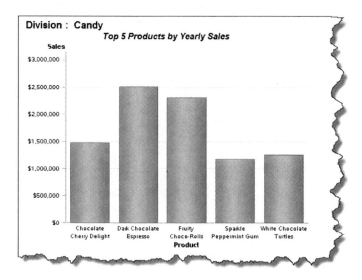

7.3.5 Working with Crosstabulation Tables

Using crosstabulation tables, you can quickly summarize relational data. In the current report layout, the crosstabulate is displaying all data by default. Because these reports have a targeted audience, you need to select the fields that highlight the data.

1. Right-click the crosstab object called Table1 and select **Assign Data** from the pop-up menu.

2. The Assign Data window appears. This window allows you to control how the table looks. You can drag and drop the items or use the **Move Items** drop-down menu while selecting the item.

In the following figure, the finished table is on the left and the Assign Data window is on the right to demonstrate how the final table appears. The columns have the vertical headings and the rows are the horizontal items. Place the data in the order in which you want it to display.

 Any items placed in the Hidden category do not display. These items are still available to use in the report.

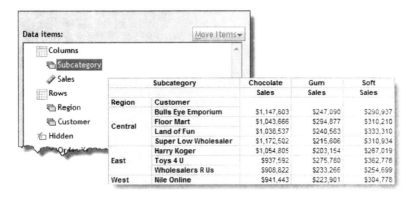

3. You can add totals and subtotals to the table. Right-click the table and then select **Total** from the pop-up menu.

The following figure shows how the table looks after the Column totals and Column subtotals are applied. The arrows indicate how the check box choice generates the totals. The subtotals provide the summary for each state and year, while the total shows the value for entire column.

If you select the row options, the totals appear horizontally.

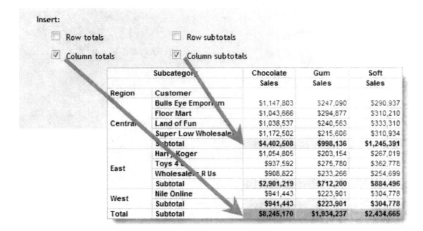

4. To create an overall report that shows the global results for the divisions, you can copy this report to a new section and then delete the group break.

    To add a new section based on the report, do the following steps:

    a. From the Table of Contents pane, select **Options > Copy Section**.

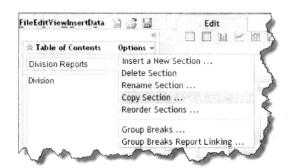

    b. Type the section name in the **Section name** field. Then select where you want the section to appear. Because this is summary data for the divisions, place this section before the division reports.

    c. This section is an exact copy of the other one but is independent of the other section. Any changes made to this section do not affect the other sections. To show the global sales, select **Options > Group Breaks** and change **Break by values of** to **None**.

There are now two sections in the report. The first section shows the overall sales for both divisions. The following figure shows all categories.

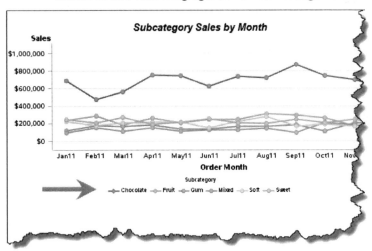

7.3.6 Creating a Template

You can create a template to use or to share with your team. Templates assist with keeping a consistent look among your department's or company's report. This is particularly important when all of your reports appear in the same area, such as a portal. The end user of the collective reports must quickly understand the reports structures so they can quickly get the needed information. When the reports have a common look and feel, this is much easier.

With the layout predetermined, it can also make report creation quicker and easier. This way the report creator does not have to focus on the format of the report, only the content. The report creator can then add the needed data and not worry about logos or footer material. Many organizations also use this area to place disclaimers or copyright information.

In the following example, you will create a template that uses dynamic text to capture the report name and the author information.

1. Select **New > Report**.

2. In Edit mode, design the layout to meet your needs. For instance, your team might decide that all reports will have a graph across the top with supporting data and analysis paragraphs at the bottom.

3. To add a custom header, click **Header** or **Footer**.

    You can add dynamic text by selecting an item from the **Dynamic Text** drop-down menu. In this example, you want a graphic with the report name and data source. For this template, the logo appears in the upper left corner. In the **Left content** tab, enter 10% for **Width** and select an image for **Content**. When you limit the width of the image, you can control where it is on the page and leave room for a title or other information.

   You can add artwork, such as logos, to SAS Web Report Studio using SAS Management Console. Refer to Section 7.6.2, "Adding Images and Logos," for more information about adding artwork files.

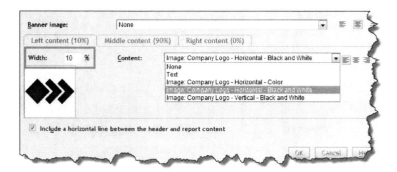

The **Middle content** tab is given the remaining width of 90%. The default report name is added.

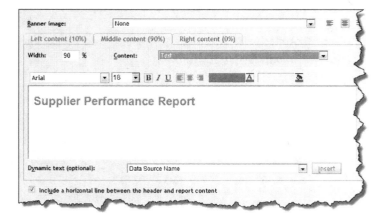

4. For the footer, type `Report Author` and select **Report Author** in the **Dynamic text** field. When viewing the report, the name of the report creator appears. This is useful in larger organizations so the report user would know whom to contact for questions.

Available dynamic text values include **Data Source Name, Data Source Description, Date the data was last refreshed, Report Author, Date Modified, Report Description,** and **Report Name,** as well as the selected value for each report prompt.

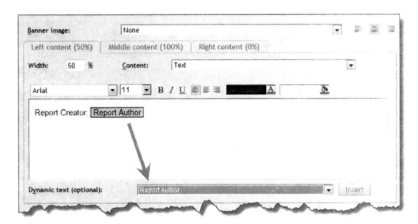

5. To save the template, select **File > Save As.**

When saving the template, select **Template** in the **Type** field. You can indicate that this template is for public use by selecting **Shared templates.**

 It is a good practice to describe the template so others understand its purpose. When there is a large list of templates, the description can better distinguish your template.

The following figure shows how the template appears when another user wants to use it to create a report. The template is available on the **Shared templates** tab.

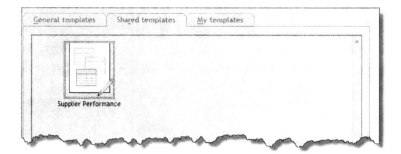

This following figure contains an example of the template when it is saved and when it is viewed. The header has the image and name, and the footer contains the report author name. When in **View** mode, the template has message boxes that indicate that no content has been assigned. When you add a data source to the report, the messages goes away.

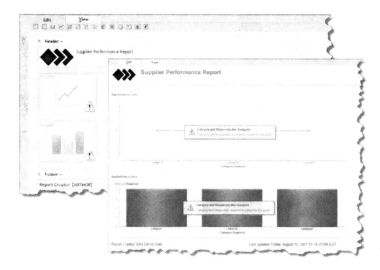

7.4 Enhancing Your Web Reports

You can change the overall appearance of your report, highlight differences in the report values, or create a custom sort order for the categories.

7.4.1 Adding Conditional Highlighting

Conditional highlighting, also, called *traffic lighting*, is easy to add to the tables or charts in your report. For the conditional highlighting, you need to establish the business rules, or otherwise indicate what is important.

In the following example, you can see how to create three conditional formats. These rules cause the following highlights to appear on the report:

- When sales percentage difference is less than the difference 12 months ago, a red down arrow appears.

- When the sales percentage difference is greater than or equal to the difference 12 month ago, a green up arrow appears.

- When the current month's sales are greater than the 3-month sales average, text appears and the background changes to green. In this case "Yes!" is the message, but you can choose any message that makes sense to your organization.

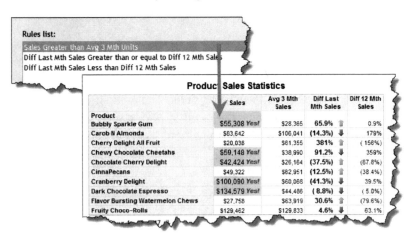

Figure 7.4-1 Using conditional highlighting

To create a conditional highlight, do the following:

1. Right-click the table where you want the filter and then select **Conditional Highlighting** from the pop-up menu. When the Conditional Formatting window appears, select **New** to create the rule.
2. On the next window, the **Rule** tab allows you to select the **Measure** and **Condition** from the drop-down menus. For **Value**, you can either type the amount or select from a list of measures.

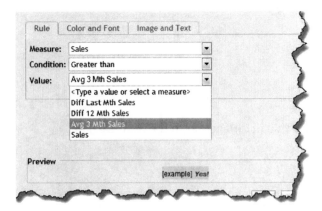

3. To add an icon to the chart, select the **Highlight by adding an image or text** check box and select the **Image** radio button. You can then select the position and icon you want to use. If you prefer to add custom text, you can also add custom text by selecting the **Text** radio button.

In the following figure, you can see an example of how to create the Text and Image elements. In the example on the left, select the **Image** radio button and the star to indicate a positive condition. There are other icons available for negative and positive conditions. In the example on the right, select the **Text** radio button and add the `Yes!` text. You might want to place other messages, based on the condition.

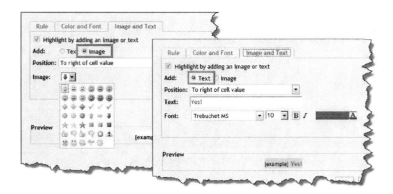

7.4.2 Creating a Custom Sort Order

By default, SAS Web Report Studio sorts category data items in ascending alphabetical order. The columns appear in alphabetical, numerical, or date order. If you need the data sorted in an alternate way, create a custom format.

This example uses the SASHELP.SHOES table to demonstrate how you can reshape the data to define a custom sort order. In the following figure, there is a before and after table showing how the sort order changes. On the left, the products appear in alphabetical order. After the custom format is applied, the products appear so that the related products are alphabetized based on product type (casual, dress) instead of gender (Men's, Women's).

Non Sorted Example		**After the Custom Sort Applied**	
Applied filters: None		Applied filters: None	

Product	Total Sales	Product	Total Sales
Boot	$2,350,543	Boot	$2,350,543
Men's Casual	$7,933,707	Men's Casual	$7,933,707
Men's Dress	$5,507,243	Women's Casual	$4,137,861
Sandal	$868,436	Men's Dress	$5,507,243
Slipper	$6,175,834	Women's Dress	$6,226,475
Sport Shoe	$651,467	Sandal	$868,436
Women's Casual	$4,137,861	Slipper	$6,175,834
Women's Dress	$6,226,475	Sport Shoe	$651,467

Figure 7.4-2 Applying a custom sort

You must have appropriate authorization to create a custom format in the BI configuration folder and the ability to create information maps. Contact your SAS administrator for more help.

To create a custom sort order, do the following:

1. In SAS Enterprise Guide, write a SAS program that creates a user-defined format that places the Product variable values in a specified numeric order. The code in the following example program creates a custom format.

 SAS Business Intelligence stores formats in a default location: <sas configuration directory>\Lev1\SASMain\SASEnvironment\SASFormats.

```
/*Create a format with the preferred sort order for the values */
libname mylib 'c:\';

proc format lib=mylib;
value fmt_product
 1= "Boot"
 2= "Men's Casual"
 3= "Women's Casual"
 4= "Men's Dress"
 5= "Women's Dress"
 6= "Sandal"
 7= "Slipper"
 8= "Sport Shoe" ;
run;

libname sasenv
 "<configuration directory>\Lev1\SASApp\SASEnvironment\SASFormats";

options fmtsearch=(sasenv);

/*register the format in the catalog */
proc catalog cat=exists.formats;
 copy out=sasenv.formats;
run;

/*Create a second version of dataset with new format */
libname sales meta library="Sales Data" metaout=data;

data sales.newshoes;
set sales.shoes;

    if PRODUCT = "Boot"            then NEWPRODUCT=1;
    if PRODUCT = "Men's Casual"    then NEWPRODUCT=2;
    if PRODUCT = "Women's Casual"  then NEWPRODUCT=3;
    if PRODUCT = "Men's Dress"     then NEWPRODUCT=4;
    if PRODUCT = "Women's Dress"   then NEWPRODUCT=5;
    if PRODUCT = "Sandal"          then NEWPRODUCT=6;
    if PRODUCT = "Slipper"         then NEWPRODUCT=7;
    if PRODUCT = "Sport Shoe"      then NEWPRODUCT=8;

format NEWPRODUCT fmt_product.;
run;
/*END OF PROGRAM */
```

Program 7.4-1 Create a custom format

2. In SAS Management Console, register NEWSHOES.

3. In SAS Information Map Studio, insert NEWSHOES into a new map.

    Instead of adding the original Product variable as a data item, use the new data item NEWPRODUCT.

    Refer to Chapter 6, "SAS Information Map Studio," for more detail on how to create information maps, add data items, and modify formats.

4. Go into the properties for NEWPRODUCT and change the type from Numeric to Category. Check the bottom of the dialog box to see the user-defined format listed.

5. In SAS Web Report Studio, select the NEWPRODUCT data item from the information map that is based on the NEWSHOES table.

6. View the report and sort the NEWPRODUCT column in ascending order. You should see the sort results based on the custom order defined by the applied format.

7.4.3 Selecting a Color Scheme

There are four default color schemes available to use with the reports. You can change the style for the current report. In Edit or View mode, select **View > Report Style**.

After you select the new style, the next time you view the report the changes are applied.

   Use the Preferences area to change the default style for all reports. Refer to Section 7.4.5, "Specify Preferences for Reports," for more information.

Figure 7.4-3

7.4.4 Customizing All Reports

You can use artwork such as your company logo or text such as disclaimer warnings in your reports. These images help your reports maintain organization branding.

7.4.4.1 Inserting Artwork into Your Report

You can add images to your reports. In some organizations, the users want to view the reports online and have the report in PDF format. Because the report has several sections, a title page and subsection title pages give the report a consistent look and match other report formats within the organization.

If you are able to save reports, you can select an image from your local machine to save with the report. However, if your organization uses the image across several publications, it might be more convenient to store the image on the server. The following example shows how to insert a title page that was previously saved as a JPG on the local machine.

To insert an image into your report, complete these steps:

1.   In Edit mode, insert a new row in the report.

2.   Click the tree icon on the horizontal toolbar and drag it to the new row. The image placeholder appears.

   The yellow icon with the exclamation mark indicates that there is not a data source for the object. In the next step, you will add the image.

3.   Right-click in the image and then select **Edit**.

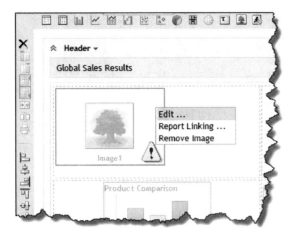

4. From the Edit Image window, you can indicate the image location and upload it to the server.

   Valid image types are TIF, JPG, and PNG.

   a. Select the **Local Machine** radio button and click the **Browse** button.

   b. Navigate to the image location and click **Open**.

   c. Click the **Select a Folder** button to select where you want to store the image on the server.

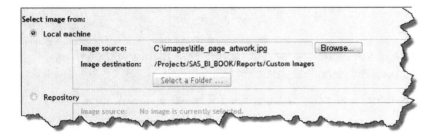

   d. You can adjust the image size if needed.

   It is a good idea to maintain the scale on images; otherwise, the final report might contain a distorted image.

5. Select **OK** to return to the report. In the following figure, you can see how the image looks in the **Edit** and **View** tabs.

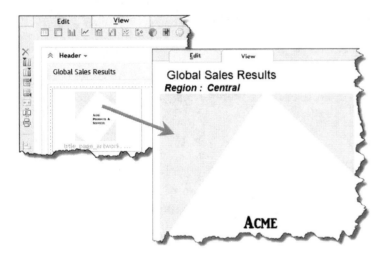

7.4.5 Specify Preferences for Reports

As you create more reports, you might find yourself in a pattern of adding the same header and footer information for each report. You can preset this information and control other settings using the Preferences window.

Use these steps to set your preferences:

1. Click **Preferences** in the right corner of the display to open the Preferences window.

2. On the **Report Creation** tab, complete any of these items to setup your preferences:

   - Select **Preferred** data source.

     Navigate to the folder that contains your data source and then select the information map, table, or cube that you want to use. The default is **Last data source used**.

   - From the **Report style** drop-down list, select the default style for creating new reports.

     The style that you select affects the color and font text of report objects such as tables and graphs. SAS Web Report Studio has four styles available: Plateau (the default), Seaside, Festival, and Meadow.

   - For the **Section header** and **Section footer**, select one or both of the following options.

     Changes made to the header and footer preferences affect all sections of a new report.

     - Banner

       Select the name of the image that you want to include in the header or footer. The list contains images that your SAS administrator has made available. If you do not want to include an image in the header or footer, then select **None**.

     - Text

       Type the static text that you want to include in the header or footer.

3. The next time you create a new report, the report uses these preferences.

7.5 Tips and Tricks

Creating standard reports typically does not address all reporting requirements. Included in this section are techniques that the authors commonly use and that require more explanation.

7.5.1 Creating Filters

Defining filters within SAS Web Report Studio allows the report developers to refine what results a final report contains. In the following example, a report is set up to filter on the year. The end user can change the filter based on the situation. Additionally, you can avoid having multiple copies of the same data or a single filter that is used by only a few people.

Do you have a filter or prompt that users continually re-create for multiple reports? Create it within one information map so everyone can use it when needed. Refer to Chapter 6, "SAS Information Map Studio," for how to create filters in the map.

To create a filter in SAS Web Report Studio, do the following:

1. In Edit Mode, select **Data > Section Filters** from the menu.

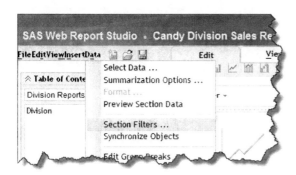

2. In the Filters window, you can create a new filter for the report, or you can use a predefined filter. In the following figure, there are two predefined filters available. For this example, you are creating a new filter. Click the **New** button to display the Create Filter window.

   In Section 7.5.3, "Linking Reports with Prompts," you will learn to use the predefined filters or prompts.

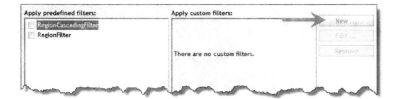

3. To create a filter for year, do the following:

   a. In the **Filter Name** field, type `Year`. This value appears in the **Filters** window.

   b. In the **Data Item** field, select the data item (Order **Year**) from the drop-down fields.

   c. In the **Operator** field, select **is after or equal to**. This filter causes one year to display at a time.

   d. Type the year in the **Type the Value to Add** field.

   e. Click **Get Values** to generate all possible values dynamically.

   To enable the **Get Values** button, refer to Chapter 6, "SAS Information Map Studio," for information about allowing dynamic values.

   f. Any added or pre-selected values can be moved to the **Selected values** area.

The following figure shows an example for a year filter set up for 1994.

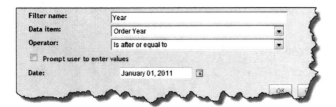

 Select the **Prompt user to enter values** check box to prompt the users for the year they want to view.

4. When you return to the Filter window, the **Year** choice is in the list. When the check box is selected, this filter is used during report generation, as shown in the following figure.

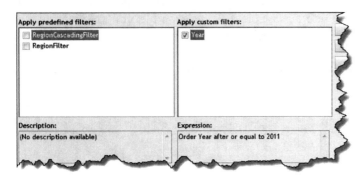

The final report has the filter applied.

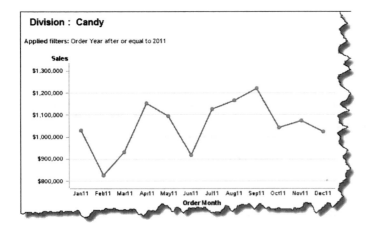

7.5.2 Using Prompts from Information Maps

When using an information map as the source, there might be prompts available. In the following example, the report you created in Section 7.3.2, "Using a Template with an Information Map," is copied to a new report and used as a foundation for the customer-based report. The cascading prompt shown in the example was created in Chapter 6, "SAS Information Map Studio."

To use a prompt in the report, do the following steps:

1. Open the report and select the **Edit** tab.

2. In the Section Data section, select **Options > Section Filters**.

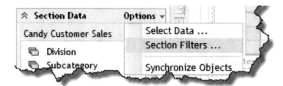

3. In the Filters windows, the **Apply predefined filters** area shows all of the prompts available for the information map. For this example, select **RegionCascadingFilter**.

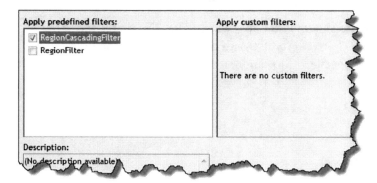

When you run the report, it prompts you for the region and customer. In the following filter, the report shows the results for the Toys 4 U customer in the East region.

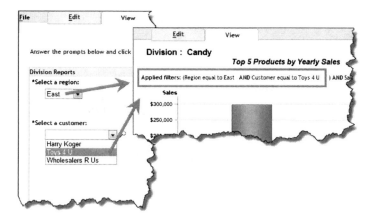

7.5.3 Linking Reports with Prompts

SAS Web Report Studio enables you to link reports together. When you link reports, you have the option to associate data item values in the source report with prompts in the target report. In this way, the prompt window bypasses the target report, and the target report is automatically subset based on the values in the source report.

In SAS Web Report Studio, you must have the capability to link reports. If you do not see the menu option, contact your SAS administrator. In addition, you can link an OLAP report to a relational report but you cannot link an OLAP report to another OLAP report.

This example shows you how to link a graph and a crosstabulation table to another report. As the input table for both the source report (Regional Sales Main) and the target report (Regional Sales Detail), this example uses a version of the Candy information map created in Chapter 6, "SAS Information Map Studio."

There are two reports you want to link, which are shown in the following figure. The first is the Show Sales (Main Report), which contains a summary chart of product sales by region. The second report is the Show Sales (Target Report), which contains specific information about the region and product.

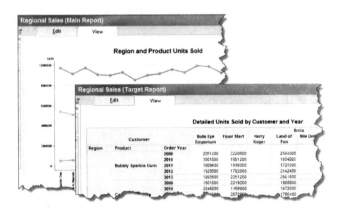

Figure 7.5-1 Reports to link

Note: Before you can create the link, the target report must exist.

1. Open the Regional Sales (Main Report) in Edit mode and right-click the object to select **Report Linking** from the pop-up menu.

2. In the Report Linking window, navigate to the target report and select it. For this example, the target report is **Regional Sales (Target Report)**.

3. To associate the Product and Region data items in the Regional Sales (Main Report) with the Product and Region prompts in target report, click the **Set Up Destination Report** button.

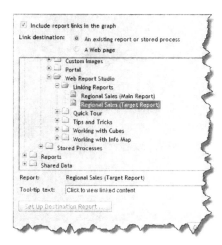

For each prompt, select the data item that matches the prompt, as shown in the following figure.

4. View the main report. In the graph, place your mouse pointer over a point. Notice that the pop-up box displays some information about the segment, as shown in the following figure.

    Click the point to send the values associated with that point (Bubbly Sparkle Gum and Central) to the Product and Region prompts for the Regional Sales (Target Report). In the following figure, you can see that the Region Sales (Target Report) shows the information you chose from the main report.

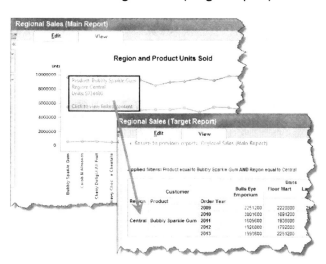

Following are a few additional tips on linking reports using prompts.

- The target report must use relational data. Many companies use an OLAP cube for the summary report and a relational table for the target report.

- If you define a link to a report that is manually refreshed, the prompt values sent from the source report are not used for the target report. Instead, SAS Web Report Studio displays the results of the last query run for the target report.

7.5.4 Adding Custom Measures

You can create simple custom measures based on the measures in the data. You can add, subtract, multiply, and create percentages in the OLAP cube or information map.

1. While in Edit mode, select **Data > Select Data**. Then select the **Custom** tab.

   For this measure, you want to see how close the actual sales were to the predicted sales figures.

2. Create a name for the measure and then build the following expression in the Expression window: [Units] * [Retail Price]

   You can type the expression or use the arrows to add measures to the **Expression** box. Type or click the operator icons to include within the **Expression** boxw. After entering your expression, click **Add**. The new custom item appears in the **Custom** item box. The following figure shows the added measurement along with the resulting report.

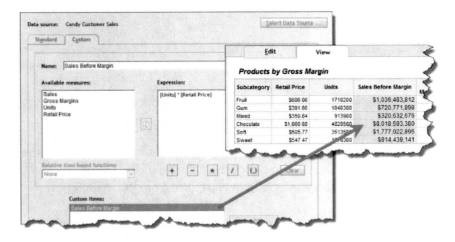

 You can add as many custom measures as you need. If you later want to remove or change the measures, use this window.

7.5.5 Scheduling Reports

For large data tables or large reports, the response time to retrieve results can be much longer than acceptable to report viewers. Reports can be scheduled to run during non-peak hours and to manually refresh when viewed, so users can quickly access results during working hours.

7.5.5.1 Creating a New Schedule

In the following example, you are going to schedule a report to run twice a day on 4 days of the week. This might not be a typical schedule for reports, but the example shows the flexibility of the scheduling.

To schedule reports, do the following:

1. In Edit mode, select **File > Schedule** to schedule the report. Save your report before scheduling it.

2. To schedule a report to run multiple days of the week at multiple times, do the following:

   a. Select **Weekly** from the **Run report** area.

   b. Select the interval from the **Weekly interval** drop-down menu. Because this report should run weekly, select **Every 1 week(s)**.

   c. Select which days you want the report to run.

   d. Click the **Multiple hours** radio button because this report needs to run every 2 hours. Select the hours of the day during which you want the report to generate. The hours display in military time.

   e. Select the date you want the report to start and end. If you select **No end date**, the report will generate until someone changes this schedule.

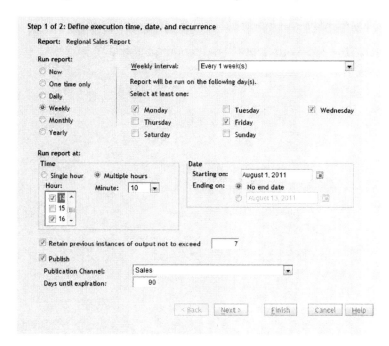

3. You can make previous instances of the output available to users. You can choose how many reports you want available. Because this is a weekly report, you should make several instances available in

case the user wants to compare a report against the previous week. Because these reports use space on your server, you might want to consider the space available to you and the usage of the report by the users.

4. Select **Finish** when you finished. A summary window outlines your choices.

7.5.5.2 Editing a Report Schedule

You can easily make changes to your report schedule at any time. Use the following instructions as a guideline when modifying a scheduled report.

5. Open the report and select **File > Schedule**.

6. Select the report you want to change and select the middle icon with the pencil.

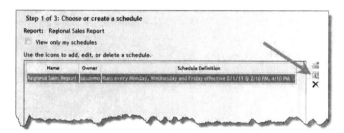

7. After the Schedule Report window appears, make the changes you want and select **OK** to save the changes.

7.5.5.3 Managing the Schedule and Report Distribution

From the Manage File window, you can quickly manage the report schedules across multiple reports.

1. Select **Manage Files** from the **File** menu on the SAS Web Report Studio home page.

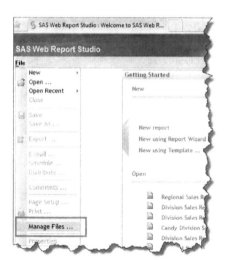

2. Click **View my scheduled and distributed reports** in the File Management pane.

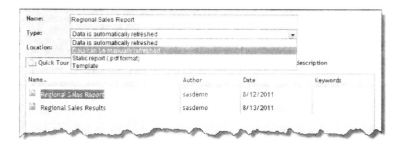

7.5.5.4 Changing a Report to Refresh Manually

When scheduling a large report, you should consider changing the default behavior to refresh manually.

On the Save As window, the drop-down box for type is available. If this is an existing report, select **Save As** to get to this window and then save the report with the same name and location to save over the existing report.

Figure 7.5-2 Controlling when reports refresh

7.5.6 Developing Reports for Wide Consumption

While developing reports, it is typical to store these in a folder structure that is not viewable by the general report-viewing audience. Some report authors place these reports in a folder under the **My Folder** structure. This ensures that no one other than the report author can view the report. However, many organizations have a team of individuals creating reports and another team that tests reports. In this structure, the site administrator might create separate folder structures with appropriate authorization levels for each group.

 Use the Export/Import wizard in SAS Management Console to migrate reports between environments, such as between a development and production server.

7.5.7 Using Group Breaks to Control Multiple Objects

You can include output from a stored process and results from an information map in the same report section and have the output from both data sources share the same grouping criteria. You perform this task by assigning the group breaks in the report to the prompts in the stored process. Combining reports from multiple sources such as information maps and stored processes provides access to more data and allows you to display various details that might not be available from one source alone.

In the following report, each object provides a different viewpoint of the data.

- The bar chart object, built from an information map, shows sales for nuts in the East region for the current year. This chart focuses on the overall product sales in the region.

- The stored process object provides a data table showing all divisions and regions sales for the past 3 years. This table focuses on the customer trends within the region.

What you want is for both objects to respond to the group breaks established for the report; when the user views the report and selects **Region**, both objects automatically filter to show that information. In the following report, the bar chart does show the correct information, but the table shows the Candy division for the Central region.

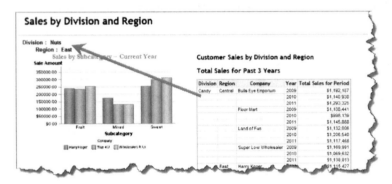

Figure 7.5-3 Stored process does not use same filter as report

To design the report so both objects display correctly, do the following:

1. In SAS Enterprise Guide, modify the stored process:

   a. Create the stored process with the prompts needed for the group breaks. The assumption is that the stored process is able to access or display data that is not available to the information map. For instance, the information map shows only the current year's data and the stored process contains historical data.

   b. Register the stored process. For this example, the stored process has Region and Division data items for prompts.

 This method requires that you have registered the stored process to execute on the logical SAS Workspace Server and to generate package results. For more information about creating stored processes with prompts, refer to Chapter 3, "SAS Stored Processes."

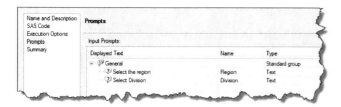

2. In SAS Web Report Studio, create a new report.

   a. Add an information map as a data source.

   b. Insert a bar chart object on the right. For this example, you want the sales amount by subcategory for each company.

   c. Insert a stored process object in the empty cell on the right.

   d. Right-click the stored process object, select **Edit**, and assign the stored process you created in step 1.

   The report looks similar to the following figure:

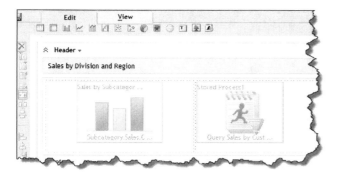

e. From the **Data** menu, select **Edit Group Breaks**.

f. In the Group Breaks window, select to break by Division and Region. Choose to display a new page for each value.

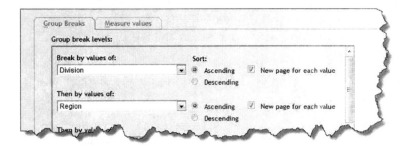

g. To associate the group breaks with the stored process prompts, right-click on the stored process and select **Assign Group Breaks**.

h. Associate the Region prompt with the Region group break and the Division prompt with the Division group break.

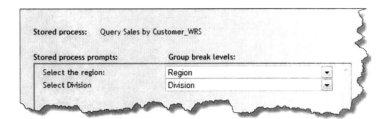

3. Click **View** to render the report. Because the group breaks provide the values for the prompts, the report bypasses the prompt window. In the following report, notice that all of the results are now based on the same criteria.

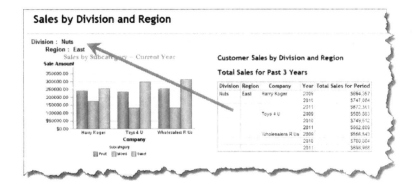

  You can assign group breaks to stored process prompts by adding compatible data items from an information map to the report. You do not have to use the data items in a table or graph. For example, if you remove the pie chart from the sample report, the group breaks can still supply the prompt values for region and product.

7.6 SAS Administrator Tasks

Using SAS Management Console, the SAS administrator can set responsibilities and make system-wide changes that assist all users.

7.6.1 Working with SAS Web Report Studio Roles and Responsibilities

To enable the availability of specific capabilities provided by SAS Web Report Studio and SAS Web Report Viewer users, each user can be assigned to one or more predefined roles. SAS Management Console has three predefined roles for SAS Web Report Studio.

Role	Purpose
Web Report Studio: Report Viewing	Allows a user to view reports.
Web Report Studio: Report Creation	Allows a user to create reports.
Web Report Studio: Advanced	Provides all capabilities in SAS Web Report Studio except the capability to manage report distribution.

The following display of the Web Report Studio: Report Creation window shows some of the capabilities initially assigned. To modify the role, such as granting the capability to access tables directly or removing the ability to access OLAP cubes, select the associated check box.

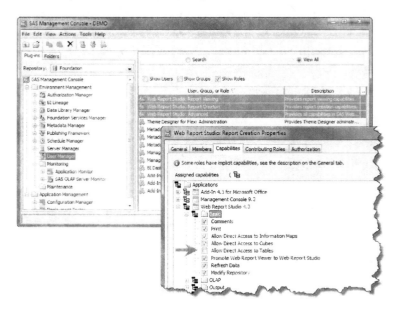

Figure 7.6-1 SAS Web Report Studio permissions

7.6.2 Adding Images and Logos

Each report that is created in SAS Web Report Studio might include one or more images. Any report can include a banner image in the header and footer of the report. Banner images make it easier for end users to identify the report and to distinguish between reports.

Banner images are stored in the following metadata location. By default, the BannerImages folder is empty.

```
SAS Folders/System/Applications/SAS Web Report Studio/Common/BannerImages
```

SAS Web Report Studio supports four image types: BMP, GIF, JPEG, and TIF.

To make a banner image or conditional highlighting image available to users of SAS Web Report Studio, follow these steps in SAS Management Console:

1. Navigate to the images folder. The default location is shown above. To display a menu, right-click on the appropriate image folder.

2. From the menu bar, select **Add Content From External File(s) or Directories**.

3. Select the file (or files) that you want to import and then click **Open**.

    Note: If you select a folder, the folder and its contents are recursively imported. In SAS Web Report Studio, banner images or conditional highlighting images that are stored in subfolders of the BannerImages and ConditionalHighlightingImages folders appear in a single drop-down list.

4. In the **Enter description** text box, enter the image name that the report builder will see in SAS Web Report Studio. Image descriptions should be fewer than 20 characters. Click **OK** to close the **Enter Description** text box.

    The imported images are available in SAS Web Report Studio after a short delay. To make the images available immediately, restart the Web application server.

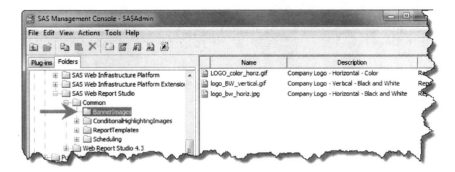

 If you modify an existing image, reimport the new image by using the preceding instructions. SAS Web Report Studio uses the updated image.

Chapter 8

SAS Add-In for Microsoft Office

Mix the Functionality of Microsoft Office
with the Power of SAS Analytics

Chapter 8

SAS Add-In for Microsoft Office

Mix the Functionality of Microsoft Office
with the Power of SAS Analytics

Microsoft Office has become a corporate standard. Nearly every organization that analyzes data has a Microsoft Excel expert who can make insightful, and even beautiful, reports that upper management loves. These users might not be SAS users but would benefit from data centrally defined within the SAS BI Server. SAS Add-In for Microsoft Office provides a bridge for these users by joining the functionality of Microsoft Office with the power of SAS analytics.

SAS Add-In for Microsoft Office extends the functionality of Microsoft Excel, Microsoft Word, and Microsoft PowerPoint by enabling you to access SAS analytics and SAS reporting functionality without any SAS programming experience. This add-in is designed for users who are familiar with these Microsoft Office applications but who might be new to SAS.

SAS Add-In for Microsoft Office is also available from Microsoft Outlook. From Outlook, you can access and share SAS Web Report Studio reports, run and review SAS Stored Process output, review dashboard indicators, and send these reports to colleagues.

In this chapter, you will learn to create reports in the various Microsoft Office applications, view the reports, and share the reports with others. Additionally there are ways to enhance your reports and some tips and tricks for improving your queries.

8.1 Getting Started

To begin learning about this tool, the following sections contain a quick review of the interface and of what you need to get started.

8.1.1 Quick Tour

In Microsoft Office 2010 and Microsoft Office 2007, SAS Add-In for Microsoft Office is available from the **SAS** tab in the Ribbon. You access the add-in from the menu options on this tab. In the following figure, you can see how the Ribbon appears in Excel.

Note: SAS Add-In for Microsoft Office is also available for earlier Microsoft Office versions.

Figure 8.1-1 SAS Add-In for Microsoft Office Ribbon in Excel

	Area	Definition
1	Insert	Use this area to access data from the SAS Metadata Server or your computer, perform tasks, view SAS Web Report Studio reports, use a SAS Stored Process, and bookmark any frequently used tasks.
2	Selection	Use this area to refresh the data in a report, modify the steps in a task, or access extra information about a selected object.
3	Tools	Use this area to send reports or graphs to other Microsoft Office applications, change the information in the spreadsheet, and access the options for the add-in. **Note**: Refer to the **HELP** button for complete information about SAS Add-In for Microsoft Office capabilities and functions.
4	Navigate	Use this area to navigate data sets with more than 500 records. You can move forward or backward through the table based on how many records you want to see at once. *This area is available only in Excel.*
5	External Data	Use this area to edit and make changes to the data on the server. *This area is available only in Excel.*

8.1.2 Prerequisites

Before you can use SAS Add-In for Microsoft Office, you must ensure that the following software is available and all necessary permissions are established:

- Microsoft Office 2007 or Microsoft Office 2010 installed on your desktop

- SAS Add-In for Microsoft Office installed on your desktop

  If the add-in is enabled in a specific Microsoft Office application, a **SAS** tab appears in the Ribbon for that Microsoft Office application.

- Permissions to access the SAS Metadata Server

Your SAS administrator sets these permissions and provides the name and location of the SAS Metadata Server. The same connections used in SAS Enterprise Guide are used.

8.2 Creating your First Report in Excel

SAS Add-In for Microsoft Office in Excel allows you to create reports with graphs and supporting data. These reports can be saved to your computer or sent to others. In this section, you are lead through the process for retrieving data from the SAS server, creating charts with supporting data, and then sharing the report with PowerPoint.

In this example, your supervisor, the regional sales manager, has requested a report that summarizes the quarterly shoe sales. Since you started working at Global Shoes Company, the management team has been discussing whether there are enough regional sales to continue to offer men's and women's shoes in all regions.

For this report, you need a bar chart that shows the sales by region for men's and women's shoes, along with supporting data in a summary report. Your supervisor requested that you create a report in Excel and a slide package in PowerPoint that she can use in an upcoming meeting.

Note: The data set for this report is shipped with the SAS Intelligence platform and is stored in a folder called SASHELP. Contact your SAS site administrator if you have difficulty finding this data set.

8.2.1 Importing Data from the SAS BI Server to Microsoft Excel

The first step to creating the report is accessing the data from the server.

Use the following steps to access the data from the server:

1. Open Excel with a blank spreadsheet. From the SAS Ribbon, select the **SAS Data** icon.

2.   When the View SAS Data window is displayed, click **Browse**.

The Open Data Source window is displayed, allowing you to explore and access data on the server or your computer. As you open each folder, all available data sets are displayed. From this window, you can select data sets defined in the metadata, information maps or OLAP cubes.

Navigate to different locations through ❶ the drop-down **Look in** menu, by selecting an icon from the menu on the left, or using ❷ **SAS Folders**. Navigate to the **SHOES** ❸ data set and click **Open**. The View SAS Data window is displayed again.

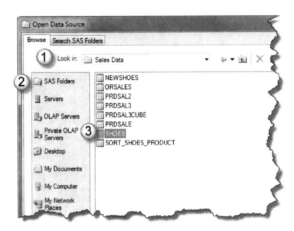

3.   Before importing the data to the worksheet, you can select how many records you want to view at once and name the worksheet.

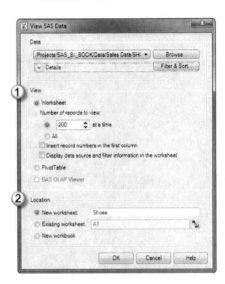

Do the following in this window to import the data:

a.   In the **View** area❶, click the **Worksheet** radio button to have all values return to the worksheet. When data is initially displayed in Excel, the view is limited to the first 500 records. If you wanted to see more records, you could increase the number here.

  You can also place data directly into an Excel pivot table by selecting the **PivotTable** radio button in the View area.

b. In the **Location** area, select **New Worksheet** and type Shoes in the field. The value used here is the name of your worksheet. Click **OK** to have the data imported to Excel.

When you return to Excel, you can see that a new tab called **Shoes** was created with the data. You can use this data with all SAS tasks and Excel functions.

If the Shoes data is updated on the server later, click the **Refresh** icon to renew the data in your spreadsheet.

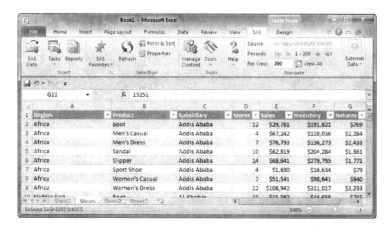

8.2.2 Filtering and Sorting Data

The SHOES data set it is not large—it has less than 500 records. Sometimes you might be working with larger data sets that take longer to import or that might have more records than you want to use. You can filter and sort data before or after importing it into Excel.

 Use the Navigation area to control how many records you view. You can also view all records.

While you can use Excel to filter and sort the records, you can sort only the records that you can see. So if you are viewing 500 records of a 10,000 table and you sort the Region column alphabetically, only the 500 records that are displayed are sorted. The remaining 9,500 records remain in the original order.

You should sort and filter the data on the SAS server prior to importing the data to Excel. You can further refine filtering and sorting by using the **Filter & Sort** button on the Ribbon.

As you continue building your report, you realize that you need regions only in the Americas and that the report is only about casual and dress shoes. In the following example, the SHOES data set is filtered to return data only from certain regions and only certain shoe types.

1. Click the **Filter & Sort** button to display the Modify Data Source window. From this window, you can preview your data, select the variable, filter the data, and sort the data. When the window is first displayed, all variables in the data set are show in the **Selected** area.

2. For this task, move **Subsidiary**, **Inventory**, and **Returns** to the **Available** area by clicking the name and clicking the single left arrow. In the **Selected** area, there should be four remaining values: **Region**, **Product**, **Stores**, and **Sales**. These are the variables you want for your report.

   Click the **Show Preview** button to see the data.

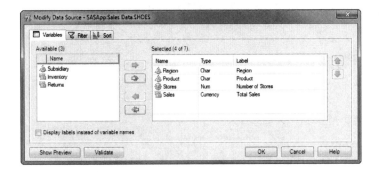

3. Click the **Filter** tab at the top of the window. This tab allows you to filter the data.

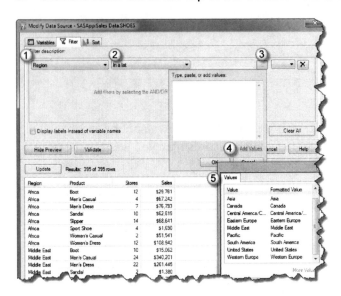

For this example, you want regions only in North America, Central America, and South America, and products specifically for men and women. Do the following to set up a filter:

a. In the **Filter description** area, click the ❶ first field and select **Region** from the drop-down list.

b. In the ❷ second field, select **In a list** for the filter method.

c. In the ❸ third field, select the **...** button at the end of the field to see the possible values.

d. When the pop-up window is displayed, click the ❹ **Add Values** text to display the values available from the table.

e. A Values window is displayed, which lists all the values for Region. Press and hold the Ctrl key while selecting the following values: **Canada**, **Central America/Caribbean**, **South America**, and **United States**.

f. When you click **OK**, the Add Values window populates. When you click **OK** again, the values are added to the field, as shown in the following figure.

4. You need to select the products you want included in the report, which are men's and women's dress and casual shoes. There are many products available, so a custom filter is useful to ensure that all values are captured.

a. To create a custom filter, click the **Advanced Edit** button.

b. When the **Advanced Edit** button is displayed, the existing query is displayed. In this case, the only filter is Region.

    Because the products you want to use all have the substring "men" in the name, you can use a combination of SAS functions and filters to create the query. The product names uses mixed case (with some values as Men's and others as Women's), so use the SAS UPCASE function to change the product name to uppercase. Then the filter looks for the character string MEN, which locates these product names: MEN or WOMEN.

c. In the **Enter a filter** area, click after the ending parenthesis and type the following:

```
AND UPCASE(Product) CONTAINS 'MEN'
```

To see a list of the SAS functions available, expand the **Functions** heading.

d. Click the **Validate** button to have SAS verify that the code is correct. A pop-up window is displayed with immediate feedback. If the expression is not valid, some helpful information to resolve the issue is provided. Click **Close** to return to the Advanced Filter Builder window.

e. You can verify that your query is going to return the data you want by clicking the **Update** button. A preview of the data is displayed. In the following display, you can see that Men's and Women's shoe types are selected for Product and that the filter returned 79 of the 395 rows in the data set.

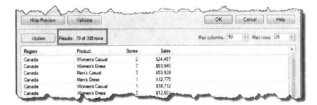

5. Click **OK** to return to the View SAS Data window. The data in the worksheet is refreshed and contains only the filtered data.

8.2.3 Building a Chart with SAS Tasks

SAS has several tasks that can be used to build reports, charts, and detailed analysis. You can also use the Excel functions with the data. An advantage to using the SAS tasks is that they do more of the work and can handle data of any size. If you use Excel to build a chart, you first have to use a pivot table or otherwise sum the data for the chart presentation.

 For SAS Enterprise Guide users, the tasks list might look familiar. SAS Enterprise Guide uses a similar set of tasks.

For the first part of this report, you want a chart that shows the sales figures for each product across each region. You can build this chart quickly using the Bar Chart task.

1. From the SHOES worksheet, click on the Region column to select the data set. Then from the SAS Ribbon, select **Tasks > Graph > Bar Chart**.

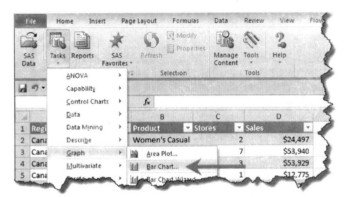

2. From the Choose Data window, ensure that **SAS Data in Excel** is selected with the SHOES data. You are creating a new report; click the **New worksheet** radio button and type `Regional Sales Report` in the field. Click **OK** to continue.

3. This report shows the total sales by region. For each region, it is important to see how each product line contributed to the overall total.

   You can select the bar chart type you want to build. SAS offers a variety of bar charts, and each icon provides an example of the chart. Do the following to customize a chart:

   Select the **Stacked Vertical Bar** icon.

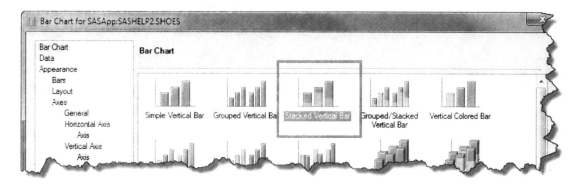

4. From the left navigation area, click **Data** to select how you want the variables used. Click and drag **Region** to **Column to chart** and then drag **Product** to **Stack**.

   Each column shows the total sales for the region, so click and drag the **Sales** to **Sum of** area.

    Click the items in the **Task Roles** area to see an explanation of how the data item is used.

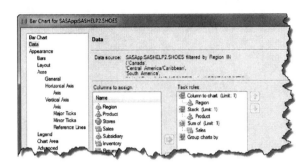

5. For the Axis title, click **Axis** under **Vertical Axis**. In the ❶ **Label** area, type `Sales by Product Line`. Then select **90** from the ❷ **Label rotation** drop-down list. The axis title will display up the side of the graph.

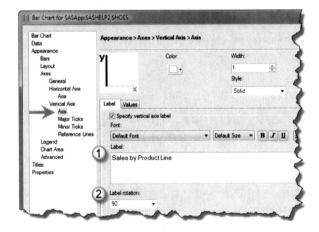

   Select **Legend** to control the legend placement and title. Under position, select **North** from the drop-down list so the legend is placed at the top of the graph.

    If you prefer to not have a legend, click the **Show legend** check box to remove the checkmark.

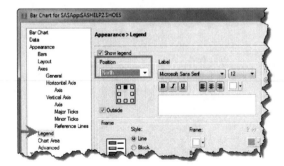

6. Click **Titles** to change the default title and footnote. Select **Graph** to change the title. Click the **Use default text** check box to clear it. Then change the title to Product Sales by Region.

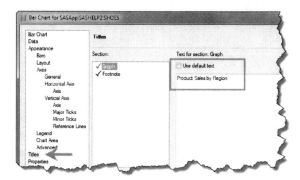

7. Click the **Run** button to render the chart in Excel.

   This is how the chart looks after it has run. Each column shows the total sales for the region broken out by the specific product types.

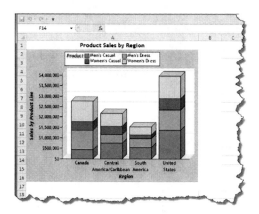

 To make changes to this graph, right click the graph and select **Modify** from the pop-up menu.

8.2.4 Building a Report with SAS Tasks

The second part of the report is a table that shows the actual sales figures. For this report, you can use the List Report wizard to show the sales by region and product, with some subtotals.

1. From the **Tasks** icon, select **Describe** and then click **List Report Wizard**. Ensure that **SAS Data in Excel** is selected and your data is shown.

You want this table on the same sheet with the chart, so click **Existing worksheet**. You can type the information in the field or use the Excel controls to navigate to the cell. Type the following text exactly as shown to ensure that the report is under the chart:

'Regional Sales Report'!B23

2. The List Report wizard starts with information about the data you are using. Click **Next**.

3. On the Define List page, all fields are available initially. You want to display only the Region, Product, and Sales fields. Select any other fields and click the **X** delete button to remove those fields. When completed, your window should match the following figure.

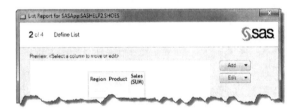

4. Click the **Edit** button and then click **Assign Columns** to display the Assign Columns window.

Click **Region** and select **Hide repeating value**s in the drop-down list. Then click and drag **Region** to the **Group by each value of** area. These actions ensure that the total displays correctly. Click **OK** to return to the Define List page.

5. Click **Sales** and select **Statistic** from the pop-up menu. You can select a variety of statistics. For this report, ensure that **Show sum value (SUM)** is selected.

6. Because you are using only one statistic, you modify the headings to remove the statistic type from them. Click the **Edit** button and select **Column Headings** from the pop-up menu. This menu allows you to control the heading names. Click the **Display the type of statistic in the column heading** check box so that it is empty. Click **OK** to continue.

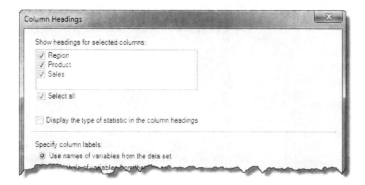

7. Click **Next** to go to step 3. Then click the **Edit** button to modify how the sales figures are totaled.

8. From the Type of Totals window, ensure that the **Grand Totals** and **Totals by Region** check boxes are selected. **Grand Totals** adds the overall total for the table and the **Totals by Regions** choice creates the subtotal rows. Click **OK** to return to the main menu.

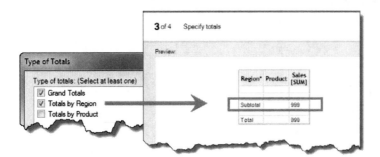

9. Click **Next** to go to Step 4. In the title field, type `Regional Sales - First Quarter Actual Sales`. Click **Finish** to build the report.

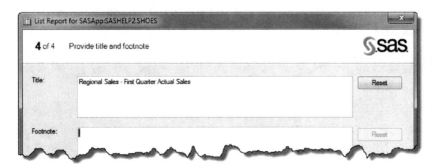

The report is displayed in Excel with the subtotal for each region and total for all regions.

8.2.5 Sharing Results with Other Office Applications

You can easily share your reports between the Office applications using the **Manage Content** icon. After you have created the information, select the **Manage Content** icon from the Ribbon. Select the reports that you want to share with Word or PowerPoint and click the **Send To** button. You can choose whether you want to open a new file or use an existing file.

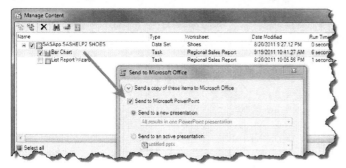

Figure 8.2-1 Sending reports to MS PowerPoint

 Click the **Allow results to be refreshed in Microsoft Office** check box to ensure that the data can be updated when in Word or PowerPoint.

Click **OK** when you are finished. The selected content is displayed in the application you chose.

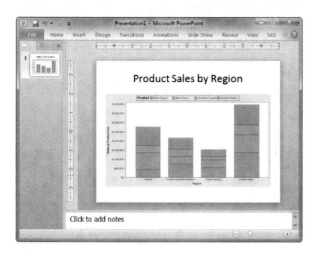

Figure 8.2-2 Reports in MS PowerPoint

8.2.6 PowerPoint and Word: Using Data to Build the Chart

One of the biggest advantages to using SAS Add-In for Microsoft Office is the ability to create a regular report once and then refresh it every week or month so that only the data changes. The formatting and other information remains constant, which means less work. In the last task, you learned how to build a report that could be shared with the other Microsoft Office applications. You can also build charts and reports in PowerPoint or Word.

In the following exercise, you will create a chart that shows the year-over-year profit for a department store. You can build this chart in PowerPoint or Word. In this example, you will use PowerPoint.

1. Open PowerPoint to a blank presentation. From the Home Ribbon, select the **New Slide** drop-down and click the **Title and Content** layout.

2. From the SAS Ribbon, select **Tasks > Graph > Line Plot**.

3. The Choose Data window is displayed. Click the **Browse** button and locate the Candy Customer Summary, which is the information map you created in Chapter 6, "SAS Information Map Studio." Click **OK** to continue.

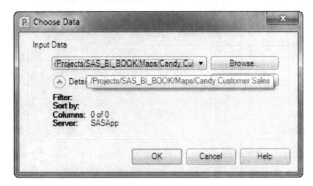

4. There are many chart choices available. For this example, you are going to do a regional analysis for each month so select the **Multiple line plots by group column** icon.

5. Click **Data** to continue. Click and drag **Order Month** to the **Horizontal** area, click and drag **Region** to the **Group** area, and then click and drag **Sale Amount** to the **Vertical** area.

   While **Sale Amount** is highlighted, click the **Summarize for each distinct horizontal value** check box. This check box totals the sales for each region so the data is displayed correctly. You need to select this check box only when the data is not summarized in advance.

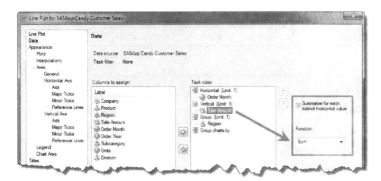

6. Click **Vertical Axis** and do the following:

   a. In the **Axis** area, type `Sales (US Dollar)` in the text box and then select **90** from the **Label Rotation** drop-down list.

b. In the **Major Ticks** area, the scale of for the vertical axis needs to be adjusted. Select the **Specify** radio button and type `0 to 1000000 by 250000` in the empty field. Click **Add** to move the value to the text box.

c. In the **Reference Lines** area, add a goal line to the chart. Click the **Use Reference lines** check box to select it. Select **Dots Dashed** from the **Style** drop-down list and select **Red** from the **Color** drop-down list.

Click the **Specify values for lines** check box to enter a target value. In the text box, type `825000` and click **Add**. The value is displayed in the text box beneath the line.

   You can have more than one reference line.

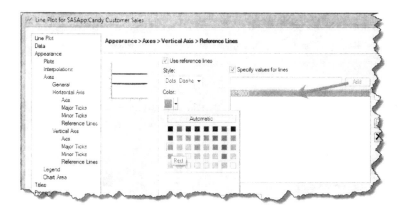

7. Click **Titles** and click **Graph**. Clear the **Use default text** check box. Then delete the text from the text box. Type `Regional Sales by Month Goal: $850,000` in the text box.

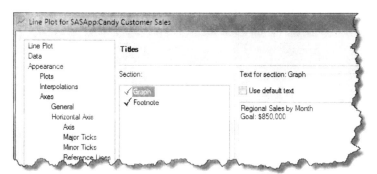

8. Click **Run** to display the chart in your presentation. If you change the way the title looks, the changes are kept even after you refresh the data. In the following figure, the title was left-justified and the goal line was changed to a smaller font.

 Click the **Refresh** icon as new data becomes available to update the values.

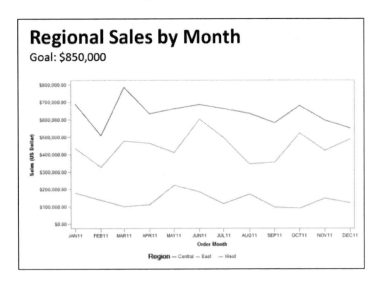

8.2.7 Accessing OLAP Cubes (Pivot Tables for SAS)

Data stored in OLAP cubes can also be accessed from Excel by using the Pivot Table wizard or through SAS OLAP Viewer. Either mechanism enables slicing, dicing, expanding, drilling, or filtering the cube structure.

The PivotTable functionality is from Microsoft; therefore, users already familiar with this interface will find viewing OLAP cubes simple. However, the new SAS OLAP Viewer provides additional capabilities, including adding bookmarks (stored to access a particular vantage point of the cube) and creating custom measures.

 SAS Enterprise Guide users will find that SAS OLAP Viewer is a similar interface to viewing cubes from SAS Enterprise Guide.

8.2.7.1 Opening an OLAP Cube

To access a cube, click **SAS Data** from the SAS Ribbon. From the View SAS Data window, click **Browse** to navigate to the cube location. After you have selected the cube, click **SAS OLAP Viewer** in the **View** area to open the cube in the SAS OLAP Viewer. You can only open SAS OLAP cubes in SAS OLAP Viewer.

Figure 8.2-3 Open data in SAS OLAP Viewer

When you open an OLAP cube, OLAP Viewer displays a table in a Microsoft Excel worksheet. OLAP Viewer functionality is similar to Microsoft's Pivot Table wizard. It allows you to move the information from columns to rows and instantly see the calculations. In addition, you can bookmark your favorite view instead of rebuilding the view each time.

Your data can be viewed as tables or graphs. The table and graph views display the current view of the cube and enable you to drill down into your data and expand or collapse levels – all in the same window.

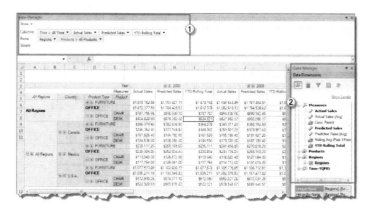

Figure 8.2-4 SAS OLAP Viewer main window

The View Manager ❶ is displayed below the Ribbon in Excel. The View Manager can display the dimensions that are being used in the columns and rows, the filters that are applied to the views, and any conditional highlights that are applied to the table view. You can customize the View Manager by changing the information that is displayed in it.

The Cube Manager ❷ allows you to interact with the cube. Using this area, you can create filters and bookmarks, and control what items are displayed.

In SAS Enterprise Guide, this tool is called OLAP Cube Explorer. For a more extensive discussion of the tool features and functionality, refer to Chapter 2, "SAS Enterprise Guide."

8.2.7.2 Adding a Graph

You can add a graph to the worksheet by selecting **Insert View** from the SAS Ribbon. You can choose the graph type, and then you are prompted for where you want to place the chart. This view of the data is synchronized with the chart data. Any changes you make to the cube view, such as expanding, drilling, or filtering, update this view as well.

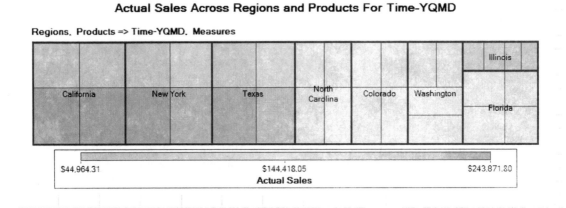

Figure 8.2-5 Auto-chart from OLAP Viewer data

 Right-click the graph and select **Graph Properties** to change the chart type or appearance.

8.2.8 Using Microsoft Outlook with SAS Reports

The new Outlook integration from SAS allows you to get reports, stored processes, and dashboards within Outlook. You can review the reports or share them with others in your organization. After opening Outlook, click **SAS** to view the SAS Ribbon.

Figure 8.2-6 SAS Add-In for Microsoft Office Ribbon in Microsoft Outlook

8.2.8.1 Using SAS Central

When you click **SAS Central**, the SAS Central folder becomes the focus. You can navigate the folders to see the SAS Web Report Studio reports that are available. When you open a folder, the reports are displayed in the work area. If you have not viewed a report before, your display will look similar to the following figure. Click **Run** to generate the report.

 You can open SAS Web Report Studio reports from the other Office applications that use SAS Add-In for Microsoft Office.

Figure 8.2-7 Using SAS Central in Outlook

After selecting **Run**, the following happens:

- A notification box appears automatically at the bottom right of the window, fading out after a few seconds.

- The report folder is shown in bold text.

- Parentheses showing the number of unread reports appear next to this bold text.

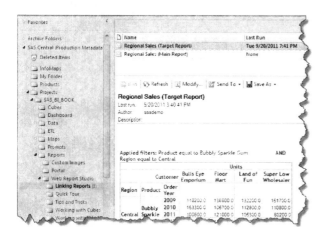

Figure 8.2-8 Viewing reports in Outlook

 When you want an update of the report, click the **Refresh** button. This task can be run in the background while you check your e-mail messages or perform other tasks.

Reports can be opened in a separate view, just as with e-mail. You can also send the report to other Microsoft Office applications to include in reports or analysis.

Figure 8.2-9 Sharing reports with other Microsoft Office applications

8.2.8.2 Sharing Reports with Others

You can send the reports to others in your organization. To send the report, right-click the report name and select **Forward** from the pop-up menu. A new e-mail is created that contains the report and a link to the report in case the receiver has issues viewing the e-mail.

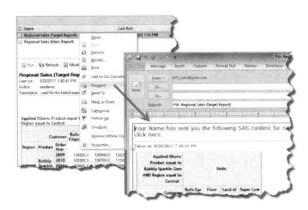

Figure 8.2-10 Sending a report in Outlook

You can add comments to the e-mail message before you send it. The e-mail includes a link to the report in case the receiver wants to get an updated copy.

 If you need to send the report to someone who does not have SAS Add-In for Microsoft Office installed, click **Save As** to create the report in PDF and HTML format.

8.2.8.3 Using the Gadget Pane

From this Gadget Pane, you can right-click a report or dashboard indicator to send an e-mail, schedule a meeting, or assign a task. A snapshot of the report item is included automatically, along with some context information. You can also refresh, copy, and modify gadgets, and view properties such as the last run time and author information.

By default, the Gadget Pane is turned off. To open the Gadget Pane, select **Gadget Pane Right or Left**. The Gadget Pane appears as another panel in Microsoft Outlook.

Using the Gadget Pane, you can monitor selected reports and dashboard indicators in Microsoft Outlook. To view the reports, click the **Report** icon. To view the dashboard indicators, click the traffic light icon.

The status bar at the bottom of each gadget displays when the report or dashboard indicator was last updated. The SAS content in the gadgets is not updated automatically. To refresh the results, click the double arrow icon.

When you refresh the results, the report results contain the current version of the data source and any values (such as format) that you specified when you first ran the report.

Figure 8.2-11 Outlook Gadget Pane

8.3 Enhancing Your Report

You can change the way your SAS data appears either using the Excel charting ability or changing the default SAS style for your report.

8.3.1 Using Excel Charts with SAS Data

You might want to use a combination of SAS data with Excel chart abilities. Excel has some extended chart rendering abilities. In this example, a SAS Task summed the number of stores and the pie chart was built using Excel. It is the best of both worlds.

The advantage to using a SAS Task is that large volumes of data can be summarized; the SAS Server can complete the task more quickly and extract only the needed data.

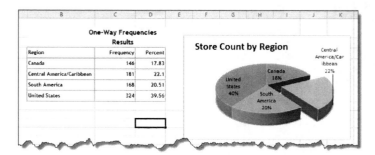

Figure 8.3-1 Using SAS data with Excel charts

8.3.2 Making It Look Nice

After you review the chart, you might want to change the appearance. If you right-click on the chart, a pop-up menu is displayed. Select **Style** to display the **Style Editor**. From here, you can change the color scheme, lines, text, or other graphic elements. SAS maintains your changes, even when you refresh the report to use updated data.

In the example, the **Style name** was changed to **Plateau** and **Scheme** was changed to **Terra**. These colors provide more contrast if the report is eventually printed in black and white.

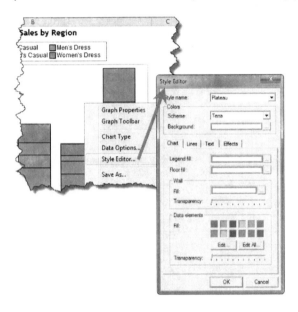

Figure 8.3-2 Using the Style Editor

Because this chart is simple, you can remove the vertical axis and display the total over each bar, as shown in the following figure. This makes it a little easier for the reader to understand immediately the overall sales value.

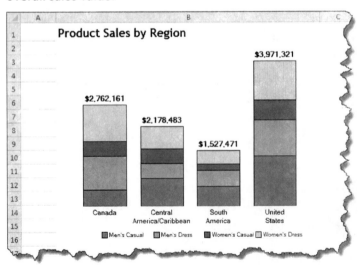

Figure 8.3.3 Modifications to chart

When working in PowerPoint, there is an extra step to accessing the **Style** menu. Right-click the item you want to modify and then click **SAS Graph v9 Object > Edit**. The graph properties and other options are available.

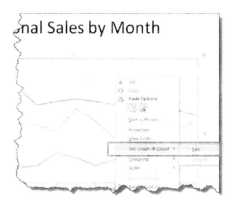

Figure 8.3-4 Editing a SAS Graph v9 Object

8.4 Tips and Tricks

The following section contains tips and tricks to assist when you are working with SAS Add-In for Microsoft Office.

8.4.1 Using SAS Functions to Build Better Filters

SAS has many built-in functions that make it easy to filter the data so you can get the exact results you want. Here are some examples of how you can use the filters in your reporting.

8.4.1.1 Working with Dates

If you plan to refresh the report and would like to include a predetermined interval of data such as data for the past day or monthly, set up a filter.

In this example, you want to see only records that were created today. Use the TODAY() function, which is a SAS function that equals todays date.

 If you are working with a date time value (i.e. 01JAN2012 00:00:00) then use the DATETIME() function instead.

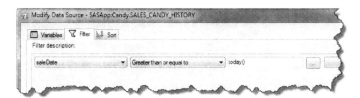

Figure 8.4-1 Working with Dates

Simple Date Filters

There are other date functions that help the report automatically display the time frame desired. For instance, sometimes you want the report to show only the records for the current year or current month. Use the YEAR() and MONTH() functions.

This advanced filter returns all orders in which the year of the order date variable is equal to the current year and the order date is equal to the current month.

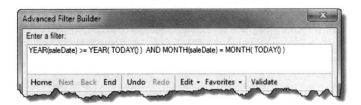

Figure 8.4-2 Working with Date Intervals

SAS has other functions for working with dates that are helpful. If you want your report to contain only the last six months of data every time it refreshes, use the INTNX function. This function increments the date based on the given interval. The following code shows the syntax for the INTNX function.

```
intnx('interval', date value, increments)
```

- *interval* can be day, month, year, week, or quarter. You can also use a time value, such as hour, minute, or second.

- *date value* is the variable that contains the date. You can also use other SAS functions such as TODAY() or DATETIME().

- *increments* is how many intervals you want to advance in the future or go back in the past. For instance, use *-2* to go back two intervals or *2* to go forward two intervals. *0* indicates the current interval.

Here are some examples using various dates:

- Filter returns orders completed within the past six months:

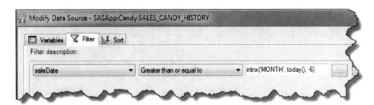

Figure 8.4-3 Filter for past 6 months of sales data

- This example is for a datetime value. Notice the datetime was added to the interval. This filter returns orders placed in the last 12 hours:

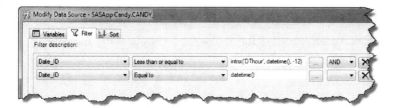

Figure 8.4-4 Filter for orders placed in last 12 hours

- Filter should return all orders closed in the current quarter:

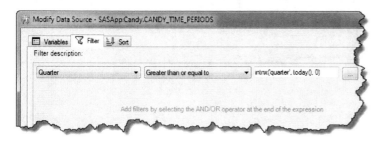

Figure 8.4-5 Filter for data from the current quarter

8.4.1.2 Working with Substrings

Sometimes when you are working with character values, you might have items that all start or end with a certain sequence. For instance, a phone number that starts with a 919 prefix or a part type that begins with X8T. The SUBSTR function allows you to isolate part of a string. The SUBSTR syntax is as follows:

substr(*variable name, start position, number of characters*)

Here are some examples of how to use the SUBSTR function.

Example filter	What the filter would find	What the filter would not find
substr(phone_number,1,3) = '919'	919-234-5555	234-919-5555
substr(part_number,10,3) = 'X8T'	ABC5-5T-YX8T	ABC5-5T-X8TY
substr(name,7,3) = 'Sue'	Betty Sue	Jamel Sute

Table 8.4-1 Using the SUBSTR function

8.4.2 Cutting and Pasting the Values

Some data filters are long lists, such as specific phone numbers or product names. The list might not have an easy pattern that you can use to filter. You can cut and paste the values into the filter.

In the following example, the orders were in the spreadsheet column. As the SAS data is loaded, the filter is created and the values are pasted into the field.

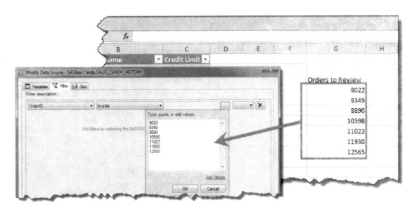

Figure 8.4-6 Cutting and pasting values

8.4.3 Copying the Modified Data to the SAS Server

With proper authorization, you can create and modify data on the SAS server using Excel. In previous releases of the SAS Add-In for Microsoft Office , you had to use SAS Enterprise Guide or SAS Management Console to add and modify data.

Use this feature when you have a smaller data set that is updated infrequently. For example, suppose you are responsible for tracking the internal training hours for each staff member. This data is used within a dashboard for the Human Resources department. However, this information is not stored by an application or in an accessible database. By saving the spreadsheet data to the server and modifying it as needed, you can manage the values within the data through SAS Add-In for Microsoft Office.

 To write back to the server, you *must* have Write and WriteMemberMetadata permissions to the library where you are going to save the data.

8.4.3.1 Creating a New Data Set on the Server

You might need to create the initial data set on the server. The following example shows how to save the data to a server location. In this example, the employee training spreadsheet is copied to the server, and it becomes a SAS data set.

1. Modify your data and then select the variable names and values in the spreadsheet that you want to copy. In the following figure, four columns (C-F) were selected. These are the only four columns needed in the new data set.

2. From the SAS Ribbon, select **Tasks> Data > Copy to SAS Server** to start the process.

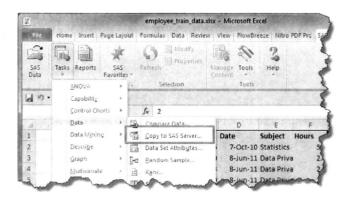

3. In the Choose Data window, ensure that the **Excel Data** field contains the entire range of data from the Excel worksheet that you want to use.

4. After you select **OK**, the SAS Task initiates.

    This window shows the data source, columns to be copied, and the new data set name and location. Your window should look similar to the following figure. Specify the destination in the Copy to SAS Server window. Click the **Browse** button to navigate to the storage area and provide the new data set name.

 Only the fields in this selected range are copied to the server.

 To replace an existing data set, select the data set and select **OK** when asked to overwrite the data. This data is permanently overwritten.

After the tasks finishes, a message is written back to the Excel datasheet. When you or a colleague want to access the data, use the **SAS Data** icon and navigate to the source on the SAS server.

8.4.3.2 Modifying Existing Data

If the administrator has given you access, you can modify data stored on the SAS BI Server. In this example, you want to add some training records to the Employee Training data set. Use the following steps to update the data and save the data to the server.

1. Open the data set from the server.

2. The SAS Ribbon has new options near the right end of the Ribbon. Click the **Begin Edit** button before you make any changes to the data set. If you make changes prior to clicking the icon, the changes are lost.

3. You can make several changes to the data:

    a. To add columns, click the **View Column Properties** button.

    b. To add records, click the **Create New Records** button.

    c. To remove records, click the **Delete Records** button.

4. After making the changes, click the **Commit** button on the SAS Ribbon to save the changes to the server.

5. Click the **End Edit** button when you have completed all edits.

8.4.4 Securing Your Information

If you are responsible for distributing a file with information, there are times when you do not want your report changed. There are a few methods you can use to prevent the data from being refreshed.

*   You can use the Excel security features that allow you to password protect the content. Refer to the Microsoft online help for more information.

*   You can break the link to the SAS data. With this method, you or others will not be able to refresh the data at any point in the future. Use the Manage Content window to break the link.

8.5 SAS Administrator Tasks

Using SAS Management Console, the SAS administrator can set responsibilities and make system-wide changes that assist all users.

8.5.1 Working with Roles and Responsibilities

To enable the availability of specific capabilities provided by SAS Add-In for Microsoft Office, each user can be assigned to one of the predefined roles. SAS Management Console has three predefined roles for SAS Add-In for Microsoft Office.

Role	Description
Advanced	Provides all capabilities in SAS Add-In for Microsoft Office.
Analysis	Provides basic data analysis, reporting, and other capabilities.
OLAP	Supports viewing OLAP cubes in PivotTables and provides other capabilities.

Table 8.5-1 SAS Add-In for Microsoft Office roles and responsibilities

The capabilities that each role allows can be further refined by the SAS administrator.

8.5.1.1 Setting Up Additional Responsibilities

The following figure of the Add-In for Microsoft Office window shows the capabilities that are assigned initially to the predefined role Advanced. This window is accessed by navigating SAS Management Console to the User Manager and selecting the **Capabilities** tab in the Add-In for Microsoft Office: Advanced window.

A user assigned to the Advanced role has the ability to save data to the SAS server. You can set this ability under the **Save or Distribute** item. If your organization does not want to allow this responsibility, then remove the checkmark.

If your organization wants to extend this ability to other roles, then ensure that this check box is selected. You can control other capabilities from this window.

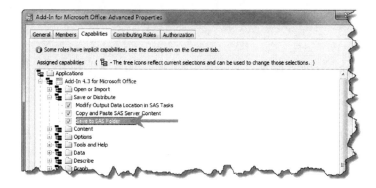

Figure 8.5-1 Setting responsibilities

8.5.2 Write or Create Metadata Access to the Library

By default, all libraries that are registered on the SAS Metadata Server are read-only. SAS Add-In for Microsoft Office users cannot create, add, or delete tables.

If SAS Add-In for Microsoft Office attempts to write data to a table on the SAS Metadata Server, (shown in Section 8.4.3, "Copying the Modified Data to the SAS Server") and the structure of the updated table is different, the metadata is not updated. For this reason, it is advisable to regulate Write access to registered tables from Excel, Word, and PowerPoint. Read access, for the purposes of querying SAS OLAP cubes using Excel, requires no special consideration.

To maintain referential integrity on the SAS Metadata Server, it is advisable to inform users of the add-in about their permissions to access registered libraries.

8.5.3 Central Storage File System

Sharing reports built within SAS Add-In for Microsoft Office requires a shared network drive. Therefore, security must be established at the physical file folder location or within the file itself.

Saving reports within a shared resource location that is not secure can run the risk of damaging the file when multiple users are interacting with and saving the report. The entire report folder should be read-only for all but the report authors; when users modify the document, they can then save it to a different location.

Chapter 9

SAS BI Dashboard

Driving Your Business from a Single Vantage Point

Chapter 9

SAS BI Dashboard

Driving Your Business from a Single Vantage Point

Reviewing pages and pages of reports eats valuable time. Typically, your executive does not spend more than a couple of minutes doing it. Missing valuable insights from data can cost an organization time and money because business decisions made based on previous experience are not always correct. A quick display of what the data history was, whether it is currently in an acceptable range, and whether it is heading in the right direction is what many executives require to analyze and improve an organization. SAS BI Dashboard offers users the ability to quickly understand summarized data, and also allows them to further research some areas.

Quick access to summarized data has always been available to SAS users through other forms, including SAS/IntrNet software, SAS Stored Processes, the Output Delivery System (ODS), and application development. However, these options require SAS programming skills and can be extremely labor intensive. The SAS BI Dashboard interface provides report creators the ability to quickly set up indicators and ranges for conditional highlighting, and to reuse these in one or more dashboards. With the 4.3 version, programmers have the ability to include custom indicators built as SAS Stored Processes.

Before beginning the work of data summarization and creating dashboards, the customer (the dashboard user) will first need to determine what exactly is needed on the dashboard. This is imperative, as simply just putting a couple of interesting indicators out there does not guarantee usefulness. There are several ways to determine appropriate dashboard content. The first and most common is the Balanced Scorecard approach. An organization's important measures are divided into four quadrants: Customers, Processes, Financial, and Employees. Each of these should include one or more objectives that are SMART (specific, measurable, attainable, relevant, and time-bound).

Another technique is to use the Goal-Question-Metric paradigm to develop the measurements to display on a dashboard. This methodology can work better for specific departments within an organization because dashboards are built solely on the goals that the organization has (or has been assigned). For each goal, questions are developed that define success for the goal, and then a set of measurements are created to answer each question in a measureable way.

The key items to note are that the dashboard must remain useful and relevant. If an indicator is always one color, ask if the range is sensitive enough or if the indicator really assists with continuous improvement. When building alerts, verify that there is an action required; if the same alert is sent every day or week, the users will likely stop listening.

After objectives or measures are defined, the report creators can begin the work of collecting and summarizing the data and creating the necessary components (data models, indicators, and ranges) to design the dashboard display.

9.1 Getting Started

As you begin learning about this tool, here is a review of the tool and what you need to get started.

9.1.1 Quick Tour

SAS BI Dashboard can be accessed directly from the Web application.

```
http://server name:port number/SASBIDashboard
```

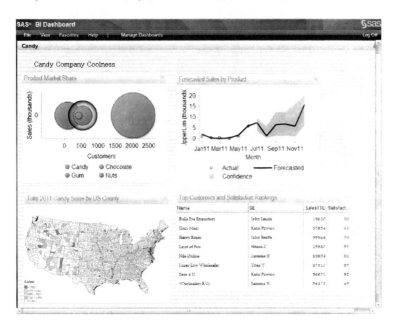

Figure 9.1-1 Overview of SAS BI Dashboard

Dashboards can also be viewed from SAS Information Delivery Portal through a SAS BI Dashboard portlet.

Figure 9.1-2 Viewing a dashboard

9.1.2 Prerequisites

Within SAS Management Console, two groups are predefined for SAS BI Dashboard users: BI Dashboard Administrators and BI Dashboard Viewers. To create content within a dashboard, you need to have your user account added to the BI Dashboard Administrators group. This group then has the BI Dashboard Administration role assigned to it. The BI Dashboard Viewers can only view a dashboard created by the BI Dashboard Administrators. They are unable to modify content.

For the purposes of the following examples, the data is summarized and already defined in metadata for the administrator. Data summarization might be required on your system and will need to be completed prior to using SAS BI Dashboard. Metadata definitions can be completed by users with WriteMetadata access, but it is typically handled by the SAS administrator in SAS Management Console. Review Section 9.3.1, "Creating Indicator Data," for other options for accessing data.

9.2 Understanding Dashboard Creation

Planning a dashboard is one of the most critical steps for success. Not only do you have to plan your data, but you also have to plan how you are going to organize the folder structure that supports the dashboard.

9.2.1 Data Summarization

Preparation of the source data is critical. In some instances, the indicator data component of the dashboard is limited to accessing only the first 1000 rows within the source table. You can decide to use only server-side filtering to get around this; however, to ensure that users have fast response times, the data should contain only the rows used in the indicator display.

Always check the number of records the data definitions returns. If they equal 1000 records, use SQL or a stored process to access the data source.

9.2.2 Organization

Folder structures used to store dashboard content within the metadata server should be planned before development begins.

Moving elements from one folder to the next or renaming the folder requires significant and tedious rework. We recommend designing and establishing the final folder structure before beginning your dashboard development work.

With one indicator data element feeding multiple indicators, or one indicator displayed on multiple dashboards, the folder structure can also help you keep everything synchronized.

You can create, find, use and administer four different types of dashboard content, dashboards, indicator data, indicators, and ranges. Other objects used by dashboards include stored processes, source data, and information maps. Security, migration, and administration of all of these items must also be considered.

Migrating dashboard content from a development server to a production server requires *identical* folder structures.

Because promotion of objects requires administrators to locate additional configuration files from the entire list of items saved in that object category, naming standards should also be implemented. Administrators will need to use a separate folder structure in SAS Management Console to select the objects for migration. This groups all of the indicator data elements together into one folder. Because object names can be saved in different folders with the same name, the administrators are unable to easily choose the components that will be migrated.

The important thing is that your organization should decide and implement one mechanism before developing any content in SAS BI Dashboard.

9.2.2.1 Folder Structure Ideas

Following are some examples of different folder structures.

Organized by group, this provides a security structure for the overall organization. Security at an individual dashboard level, however, is difficult.	All objects for a dashboard are stored together, allowing for security to be applied at a project and a dashboard level.
Organization Name	Project Name
<ul><li>Dashboard1</li><li>Dashboard2</li><li>Indicator Data<ul><li>DB01 _1</li><li>DB01 _2</li><li>DB02 _1</li></ul></li><li>Range<ul><li>DB01 _1</li><li>DB02 _1</li><li>DB02 _2</li></ul></li><li>Indicator<ul><li>DB01 _1</li><li>DB01 _2</li><li>DB02 _1</li></ul></li></ul>	<ul><li>Dashboard1<ul><li>Dashboard1</li><li>DB01_Data_1</li><li>DB01_Data_2</li><li>DB01_Range_1</li><li>DB01_KPI_1</li><li>DB01_KPI_2</li></ul></li><li>Dashboard2<ul><li>Dashboard2</li><li>DB02_Data_1</li><li>DB02_Range_1</li><li>DB02_Range_2</li><li>DB02_KPI_1</li></ul></li></ul>

Table 9.2-1 Folder structure ideas

9.2.2.2 Naming Convention Ideas

In the folder structure examples above, the file objects are following a naming convention as well. This file name is viewable only by dashboard developers and the end users see only the title given to the element.

```
Database Number _ Object Type _ Business Name
```

It is recommended that dashboard developers implement a numbering scheme because it assists SAS administrators in quickly finding the objects in SAS Management Console. As shown in the following figure, an administrator can then quickly select specific dashboard elements for partial promotion from a development server to a production server.

Figure 9.2-1 Content list

Including the type of object in the name is an extremely useful method of naming the objects.

9.3 Creating Your First Dashboard

It is important to understand the relationships between the various components. There are four components that make up a dashboard within SAS BI Dashboard: data models, ranges, indicators, and dashboards.

Depending on the type of indicator, both indicator data and ranges must be defined.

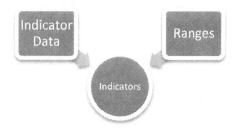

Figure 9.3-1 Creating indicators

 Reusing data models and range definitions for multiple indicators is an option; however, remember that modifications to either component affect all indicators that use it.

Multiple indicators can reside on a dashboard.

The data model and range definition can be completed during the creation of a new indicator, but for the remainder of this topic, each component is developed independently.

To create new content, either use the file menu or use the Create New Content menu located in the center of the page.

Figure 9.3-2 Creating new content using a new content page

Figure 9.3-3 Creating new content with the menu

9.3.1 Creating Indicator Data

Data can be accessed directly using SQL code, retrieved from metadata definitions and information maps, or derived from stored process code.

After selecting the menu to create new indicator data, give the object a name and arrive at the interface to define the data source. The list of options is below; select the appropriate source.

Data Source Type	Description
Information map	A map, defined in SAS Information Map Studio
SQL query	SQL code that accesses a data source
Stored process	A stored process written to generate a package result of SAS data in a specific format consumable by the dashboard
Table	A data table defined in metadata

Table 9.3-1 Object names

SQL queries include only the select program statement and can point to any library that is defined and available to the workspace server.

For this example, select the information map Customer and Product Data (which you built in Chapter 2, "SAS Information Map Studio") to create indicator data for use by other elements in the dashboard.

1. Click **Browse**.

2. Select the Information Map and click **OK**.

3. Review and validate the information within the **Data Mapping** and **Query Results** tabs in the **Preview Design** area.

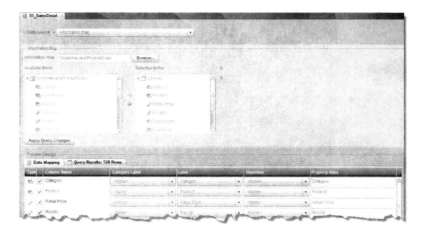

9.3.2 Creating a Range

Range definitions are used by most indicators to highlight when your results are on target. Creating ranges is technologically simple; the real work is defining values that effectively highlight issues for your organization to take appropriate action.

The following steps detail how to create a range for sales totals. In this example, the candy company has agreed that any sales total under 100 is below target, anything over 200 is above target, and everything in between is on target.

1. After selecting the menu to create new range, give the object a name and arrive at the interface to define the range.

2. Select the **Add Interval** button.

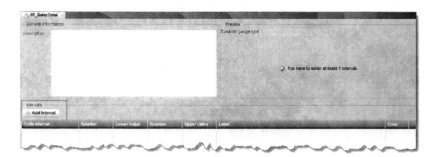

3. Type in the range value of 100.

4. Do steps 2 and 3 again with the range value of 200.

5. According to the initial range definition, all values are now red.

6. Use the drop down menus for **Code Interval** to change to: **Below Target**, **On Target**, and **Above Target**.

    Note that the colors have changed automatically to red, green, or yellow for the interval types.

7. Change the yellow color by double-clicking on the yellow and choose blue.

8. Now select the **Save** icon at the top or click **File>Save** from the menu and save the range with the appropriate name and folder.

9.3.3 Creating an Indicator

After developing the required indicator data and range, it is time to create an indicator. There are various display types available. When using stored process, code you have an essentially limitless toolkit of display possibilities.

Predefined indicators include the following:

Bar Chart with Bullet	Interactive Summary and Bar Chart	Scatter Histogram
Bar Chart with Reference Lines		Scatter Plot
Bubble Plot	Interactive Summary and Scatter Plot	Schedule Chart
Chart with Slider Prompt		Simple Bar Chart
Clustered Bar Chart	Interactive Summary and Targeted Bar Chart	Spark Table
Custom Graph	Key Performance Indicator (KPI)	Stacked Bar Chart
Dual Line Chart	Line Chart with Reference Lines	Targeted Bar Chart
Dynamic Prompt	Needle Plot	Tile Chart
Dynamic Text	Pie Chart	Vector Plot
Forecast Chart	Range Map	

In this example, the candy company is interested in viewing the market share of their products. To represent this, the developer has decided on a tile chart.

1. Create a new indicator.

The resulting screen shows that the tile chart display type has been chosen, as well as the indicator data you created earlier.

On the right side of the screen, the indicator properties are displayed.

2. Move to the **Role Mapping** area to select the data elements.

3. Set the following values:

    a.   **Range Value** = **Retail Price**

    b.   **Tile by** = **Product**

    c.   **Tile size** = **Retail Price**

The indicator automatically refreshes to display the result.

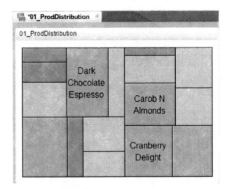

4.   Now select the **Save** icon at the top, or click **File>Save** from the menu and save the indicator with the appropriate name and folder.

9.3.4 Arranging the Dashboard

Once indicators are defined, you can add the indicators into the dashboard and arrange them within the dashboard layout screen.

1.   Select the new dashboard.

2.   Drag indicators from the Library to the Dashboard.

3.   Update the indicator box name by clicking the indicator once and modifying the **Object Name** area in the Properties pane.

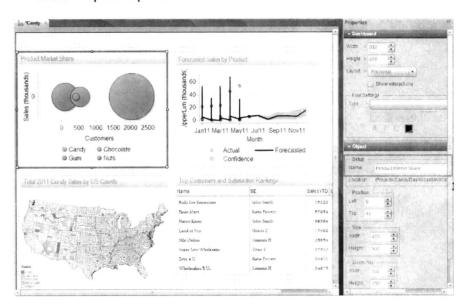

9.3.5 Alert Framework

Dashboards are considered *pull* environments, where users are tasked with reviewing and analyzing the reports on their own. However, there are times when information must be *pushed* to users through alerts. SAS BI Dashboard includes an alert framework to review events and send alerts to users when conditions have occurred. Alert definitions can either be defined globally for a group of users, or it can be handled by the individual dashboard users so that they can define when and what alerts to receive. However, as a dashboard creator, you need to enable personal alerts.

The following topic details how, as a dashboard developer, to enable the indicator for an alert, and what the users must do to create and manage their personal alerts.

9.3.5.1 Enabling Personal Alerts for an Indicator

At the bottom of the properties for an indicator is the **User Personalization** area. For dashboard users to set their own alerts, you must select the **Manage Alerts** check box.

Figure 9.3-4 Enabling personal alerts

9.3.5.2 Setting Up a Personal Alert

When viewing the dashboard, each indicator has a drop-down menu that includes options to write comments, print, or set personal alerts.

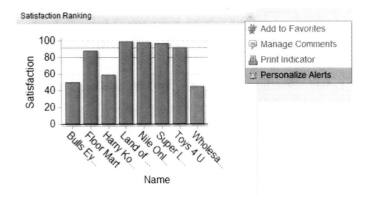

Figure 9.3-5 Personalize alerts using the Dashboard Indicator menu

After selecting **Personalize Alerts,** a pop-up window appears, which contains the window to create a new indicator alert.

Figure 9.3-6 Editing personal indicator alerts

You must update the **Alert Settings** section with a name, a gauge, and when to trigger the alert. Within the **Delivery Method** section, you can choose to display the alert within an alert portlet from SAS Information Delivery Portal or to receive an e-mail.

After entering in everything, do not forget to select the **Save Alert** icon at the top of the screen.

Figure 9.3-7 Saving new indicator alerts

9.3.5.3 Creating a Global Alert

When creating a new indicator, the bell icon is available at the top of the Properties pane. Select this bell to generate global alerts.

Figure 9.3-8 Create a global alert

The following display allows the dashboard developer to create multiple alerts for the indicator.

Figure 9.3-9 Create multiple alerts

Alert settings include the name, the gauge, what triggers the alert, and when the alert is sent.

The delivery method allows a choice between viewing alerts through an Alerts portlet (viewable in SAS Information Delivery Portal) or by receiving an e-mail.

   Use of alert emails requires that email addresses are defined in SAS Management Console for the user receiving the alert message.

1. Global alerts can be emailed to a distribution list. To set the list of alert subscribers, scroll to the right of the delivery method section of the Edit Indicator Alerts screen to see the green plus sign (+) icon. Click this icon to open the **Add Users and Groups** screen.

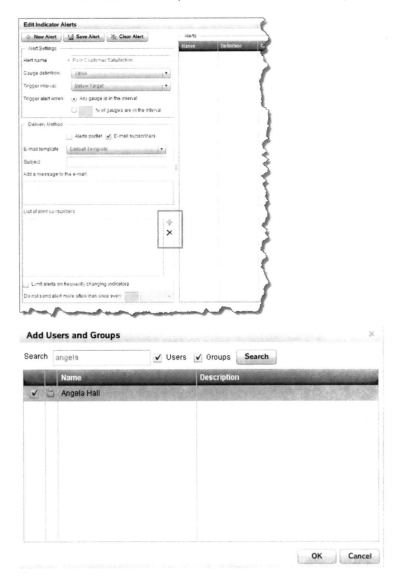

2. After adding users and groups, be sure to select the **Save Alert** button on the top of the Edit Indicator Alerts window to save all the changes.

9.4 Enhancing Your Dashboard

SAS BI Dashboard has many features that allow you to customize it, such as changing the fonts and colors schemes. Stored processes can even be used to provide an additional, powerful layer of functionality.

9.4.1 Font Styles and Sizes

While modifying the layout of your dashboard, you can control the fonts across all indicators. This is handled at the dashboard level so the view remains consistent.

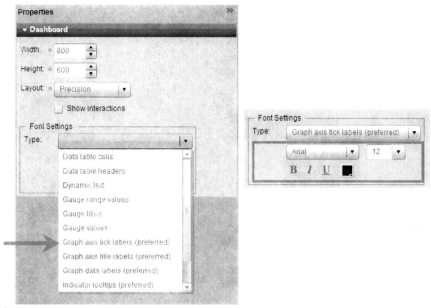

Figure 9.4-1

You can modify the following different types of text:

Data table cells	Graph axis tick labels (preferred)
Data table headers	Graph axis title labels (preferred)
Dynamic text	Graph data labels (preferred)
Gauge range values	Indicator tooltips (preferred)
Gauge titles	Panel titles
Gauge values	Prompt items (preferred)

9.4.2 Resizing

The final dashboard view for users depends of display settings in several locations. Each indicator has a width and height setting. Gages have additional sizing settings. The dashboard and the dashboard portlet, viewed from the SAS Information Delivery Portal, also have settings.

The Properties pane lists the sizing options when they are available.

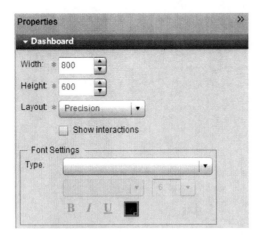

Figure 9.4-2 Dashboard properties

Figure 9.4-3 Dashboard portlet properties

Following is an example of a portal page on which the dashboard is not viewable in the entire portlet. The Sales Detail Report is running off the right side of the page, and the scroll bar is not large enough to see the last column.

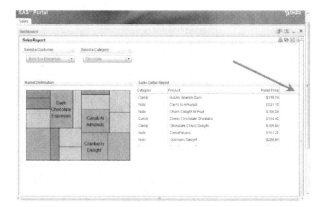

Figure 9.4-4 Portlet clips dashboard content

The first step is to edit the dashboard. Immediately, you can see that when the objects were added to the dashboard, they were not properly sized. The Sales Detail Report runs off the grid lines for the dashboard size.

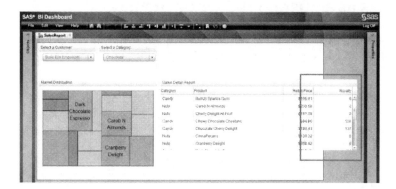

Figure 9.4-5 Dashboard object is too large

Within the dashboard edit window, the object is sized smaller and the column widths are adjusted by clicking and dragging elements.

Sales Detail Report			
Category	Product	Retail Price	Royalty
Candy	Bubbly Sparkle Gum	$105.91	0
Nuts	Carob N Almonds	$230.58	0
Nuts	Cherry Delight All Fruit	$157.38	0
Candy	Chewy Chocolate Cheetahs	$84.80	106
Candy	Chocolate Cherry Delight	$190.43	137
Nuts	CinnaPecans	$139.32	0
Nuts	Cranberry Delight	$258.42	0

Figure 9.4-6 Resize the dashboard object

9.4.3 Color Schemes

Colors are defined within the range definition. The following indicators use these colors. Colors change based on the measure data values.

Bar Chart with Bullet Display	Interactive Summary and Targeted Bar Chart Display	Spark Table Display
Bar Chart with Reference Line Display	KPI Display	Targeted Bar Chart Display
Combo box with Gauge	Dynamic Gauges	Tile Chart Display
Dynamic Text Display (Optional)	Line Chart with Reference Lines (only for Reference Lines)	Vector Plot Display
Interactive Summary and Bar Chart Display	Range Map Display	Waterfall Chart Display
Interactive Summary and Scatter Chart Display	Scatter Plot Display	

If bar colors based on categorical values need to be defined, use the stored process display to define and lay out the chart according to your color specifications.

9.4.4 Using a SAS Stored Process as a Custom Indicator Display

If the required indicator is not available in the product itself, you can code the graphical output within a SAS Stored Process. Refer to Chapter 3, "SAS Stored Processes," for information on how to create a stored process for consumption by a dashboard.

Once the stored process is completed, generate the full URL path to execute the stored process in a Web browser. An example is:

```
http://server name: port number/SASStoredProcess/do?_action=execute&_program=/Projects/
Candy/Dashboard/Indicator/GMAP_Indicator
```

This is copied into the indicator properties and will be used by a dashboard to generate the display for the users. The machine and port information are required; therefore, administrators must be aware that the properties require modification during promotion.

1. Within SAS BI Dashboard, select **New Indicator** and choose **Custom Graph** as the display type. The indicator data and range are not required.

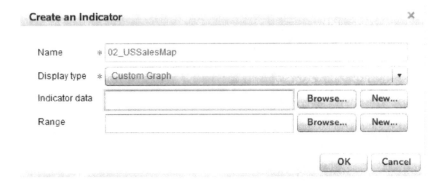

2. Within the **Role Mapping** area of the Properties pane, copy the stored process URL.

3. The stored process executes and provides you with a preview.

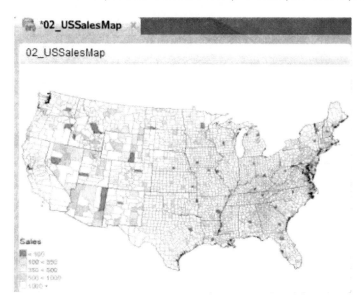

4. Now select the **Save** icon at the top or click **File>Save** from the menu and save the indicator with the appropriate name and folder.

9.4.5 Using a SAS Stored Process as Indicator Data

When data must be refreshed immediately or when the table does not already exist within the metadata, the stored process provides another data source for the dashboard. The stored process must use the SAS Publishing Framework to generate a package on the SAS Content Server.

Following is a sample stored process that generates the number of units sold over the past three months.

```
/*================================================================*/
/*Set the metadata library location*/
libname candy meta library="Candy";

/*Create the work data table for use by the dashboard*/

proc sql;
create table result
as select distinct a.product, sum(b.units) as qtrly_total
from candy.candy_products as a
 inner join candy.candy_sales_history as b
on a.prodid = b.prodid
where b.date >= intnx('month', today(), -3)
group by a.product;
 quit;
/*Grab the location of the current session's work folder to place the packaged data*/
%let temp_path=%sysfunc(pathname(work));
/*Output the package*/
data null;
length path $32767;
rc = 0;
pid = 0 ;
description = 0;
name = '';
call package_begin( pid, description, name, rc);
/*result = the name of the data table*/
/*"Last 3 month sales total" = the description of the package*/

call insert_dataset( pid, "WORK", "result", "Last 3 month sales total", '', rc);

/*Creates a package in c:\sas\packages called ThreeMonthTotal"*/

call package_publish( pid, "TO_ARCHIVE", rc, "archive_path, archive_name,
archive_fullpath","&temp_path", "ThreeMonthTotal", path );
call symput( '_ARCHIVE_FULLPATH', path);
call package_end( pid, rc);
run;

/*================================================================*/
```

Program 9.4-1 Stored process for custom indicators

When saving the code as a stored process, the following items must be done for a dashboard to use the package.

1. Do not include %STPBEGIN and %STPEND in the code. Always click **No** to ensure that these macros are not appended automatically.

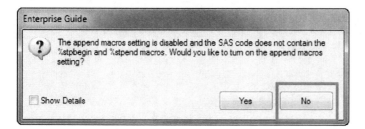

2. When defining the stored process, select **Package** as the result type.

3. Select the SAS Stored Process Server.

Within SAS BI Dashboard, create the indicator data by selecting the stored process option for the data source. Browse to the new stored process and wait for the data set name to resolve. Click the **Query Results: 16 Rows** tab to view the output.

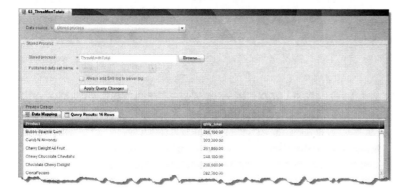

9.5 Tips and Tricks

You can easily provide more depth to your dashboard by creating drill paths that link to other reports in the system, prompts that allow the users to prefilter the data they want to see, and relationships between indicators. The following tips show how to add these features.

9.5.1 Creating Drill Paths to More Detail from Indicators

Dashboards are meant to act as a starting point for managers to use in interpreting current status. When indicators require investigation, users immediately ask for the ability to drill down into more detail.

From the Indicator Properties pane, select the link icon to set up indicator links.

Figure 9.5-1 Establish indicator links

The Set Up Link window appears, allowing you to create a connection between the current indicator and another object.

Figure 9.5-2 Using the Set Up Link window

Available link types include those in the following list. In the examples in this section, we will touch on the use of links in Web reports and stored processes because they are more commonly used.

Dashboard	SAS Information Map
External Link	SAS Stored Process
Indicator	Web Report
Portal Page	

9.5.1.1 Linking to a SAS Web Report Studio Report

The candy organization is interested in going from a basic report of top customers and their customer satisfaction rankings to a Web report showing what sales those customers have generated over the last several years. The Web report was already built and has a prompt for the customer name, so we only need to link between the dashboard indicator and the web report.

1. After opening the Set Up Link window as defined in Section 9.5.1, "Creating Drill Paths to More Detail from Link Indicators," select **Web Report** as the link type.

2. Click the **Browse** button and navigate to the Web report.

3. If the Web report has a user prompt, you can link the parameter name and indicator data value by selecting the green + button.

    The parameter name for Web reports is the prompt name. It *must* exactly equal the value of the displayed text in the SAS Web Report Studio prompt. The following figures show where you can grab this parameter.

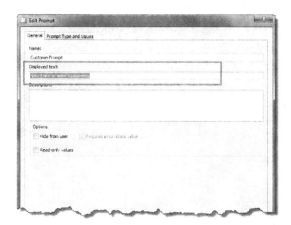

Figure 9.5-3 Displayed text within a SAS Information Map Studio prompt

Figure 9.5-4 Prompt displayed text from SAS Web Report Studio

The same text is then typed into the **Displayed Text** field on the Set Up Link screen, as seen below.

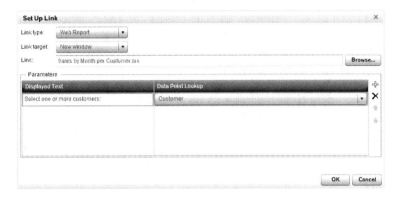

Figure 9.5-5 Establishing the link from a dashboard to a prompted Web report

 If the parameter is not successfully passing from the dashboard to the Web report, verify that there are no extra spaces, the case completely matches, and the text is exactly the same.

9.5.1.2 Linking to a Stored Process

The candy organization has asked if they can open the stored process that creates custom-order print outs from the spark table within the sales report dashboard. You created the custom order print out in Chapter 8, "SAS Stored Processes." The parameter was order_number.

1. After opening the Set Up Link window as defined in Section 9.5.1, "Creating Drill Paths to More Detail from Link Indicators," select **SAS Stored Process** as the link type.

2. Click the **Browse** button and navigate to the stored process.

3. If the stored process has a parameter prompt, you can include the parameter name and the indicator data value by selecting the green + button.

The final window then provides all the information needed to link between the two components.

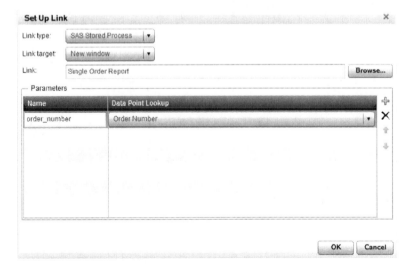

9.5.2 Cascading Prompts

Developers will find that users request more and more information to be available immediately on the dashboard window. Prompting the user first for filter information and displaying related data within the indicators can provide a flexible environment to help address this.

The candy company sales team requires a sales data dashboard organized by customer and product categories to better understand which products are most popular, by customer.

Following are the steps to create a dashboard with these cascading prompts.

1. Within SAS Information Map Studio, create (or update) an information map with prompted filters. The following two prompts must have default values assigned.

    -   Customer

    -   Customer and Subcategory

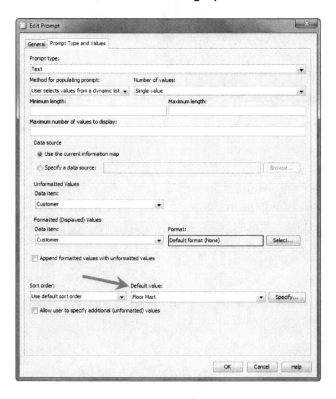

2. Within SAS BI Dashboard: Create three separate indicator data objects.

    - Unfiltered

    - Customer Filtered

    - Customer and Subcategory Filtered

    When creating the filtered indicator data objects, ensure that the corresponding filter that you built in step 1 is added to the object.

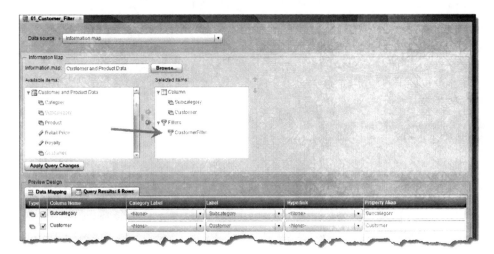

 If you have trouble adding and saving the filters in the indicator data, it is because you have forgotten to add default values to the information map prompts.

3. Within SAS BI Dashboard, create two dynamic prompt indicators. Select New Indicator from the menu and choose the Dynamic Prompt indicator type.

    a. Customer

        Because this is the first prompt, you want to have all customers within the list. Therefore, select the unfiltered indicator data and set **Role Mapping** to use **Customer**.

    b. Customer and Subcategory

        This prompt will list all subcategories that exist for a particular customer. You need to use the customer filtered indicator data and select **Subcategory** within the **Role Mapping** area.

 All other indicator types can be used as the source for interactions, except dynamic text and interactive displays.

4. Add these objects to the dashboard.

5. Define the interactions.

   a. Select the **Customer Dynamic Prompt** indicator. Choose all the other objects, select **Server-side filter** and match them with the customer and customer prompt.

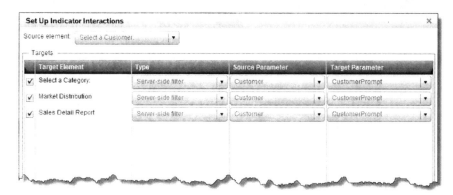

   b. Select the **Subcategory Dynamic Prompt** indicator. Choose the two graphics indicators, select **Server-side filter**, and select subcategory for **Source Parameter** and subcategory prompt from **Target Parameter**. The **Target Parameter** column will list the prompts defined in the information map. Currently this is not a distinct list of values, but a full list of all prompts available in all information maps used in the dashboard.

   While making **Source Parameter** and **Target Parameter** selections, choose any one of the available values that corresponds to the correct parameter name. In the current release, the retrieved text for this prompt contains duplicates.

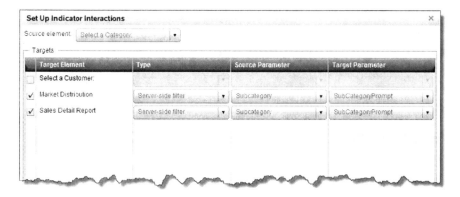

6. Save the dashboard. Test the prompts within the dashboard viewer. The following two figures show the different results depending on the prompt selections.

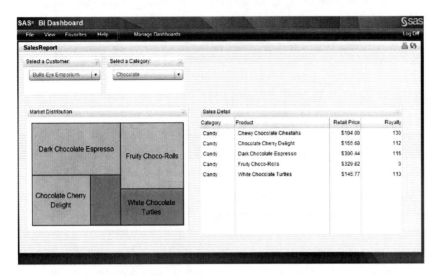

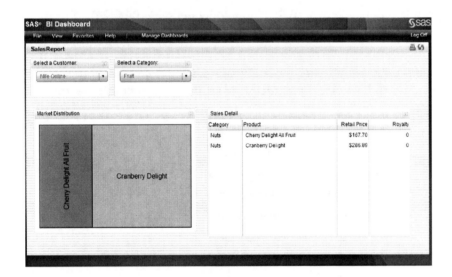

9.5.3 Client-Side versus Server-Side Filtering

When defining interactions between filters and indicators, both client and server-side filtering options are available. Client-side filtering is applied to the data available within the browser session. Server-side filtering is applied against the data on the server before it is rendered on the client.

Most indicators can accept the filter, based on your selection. However, the KPI and Interactive displays options cannot act as the target element when using client-side filtering. Similarly, dynamic text and interactive displays cannot act as target elements for server-side filtering.

Server-side filters can use indicator data built only from information maps or stored processes, while client-side can be run on any indicator data. Included below is more detail on what server-side and client –side filtering support.

	Server-Side	**Client-Side**
Source Indicators	All except dynamic text and interactive displays	
Target Indicators	All except dynamic text and interactive displays	All except KPI and Interactive displays
Indicator Data Sources	Only Information Maps and Stored Processes	All
Data Notes	Prompted filters are required for each parameter and a default value is required.	Do not specify a default value for the mapped parameter so that all information is available to user while interacting with indicator.
Query Action	Server	Client
Performance Notes	All queries execute on the server, therefore response time could increase while communicating across system.	All queries execute within the client browser window. Time to load the initial dashboard view could increase, however all subsequent interactions take place immediately.

Table 9.5-1 Server-side and Client-side Filtering Matrix

 If you decide to switch between server-side and client-side filtering, ensure that the dashboard is closed before modifying the default prompt values and saving the information map.

9.5.4 Brush Interactions between Indicators

Creating interactions between indicators provides users more information and flexibility to analyze relationships between the results. For this example, the candy company really likes the sales report dashboard. However, the product names are not displayed in all instances in the range map. They would like a detail report available next to the range map where users can click on an area in the map and view the corresponding row in the data table.

After creating a spark table using the same indicator data that the range map is built from, the interactions are defined by clicking once on the first indicator (the range map) and choosing the set up interactions icon from the dashboard menu bar (indicated in red in the following figure).

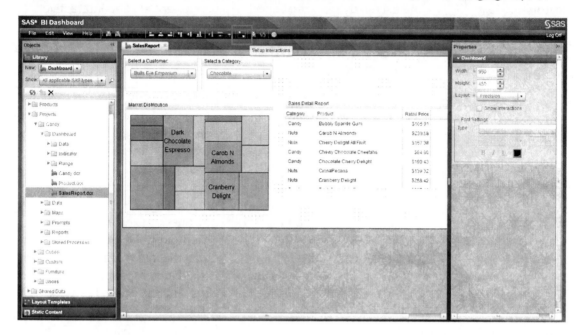

Figure 9.5-6 Set up interaction icons

In the subsequent window, select the spark table (already named Sales Detail Report), choose the Brush type, and then select the source and target parameters. The parameter values represent which content information to use to join the two indicators together. Because both indicators are built from the same indicator data, these parameters are identical. When indicators are built from different data sources, this type of prompt allows you to select different columns to match data.

Figure 9.5-7 Choose the indicator interactions

After saving and viewing the dashboard within the dashboard viewer, you can click an area of the range map ❶, the corresponding record will then be highlighted with a different color in the spark table ❷.

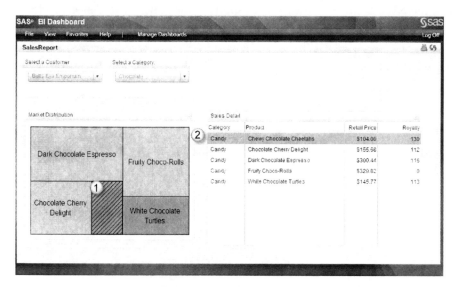

Figure 9.5-8 First interaction example

Included is a second example where the user has selected Watermelon Taffy ❶ and the record in the spark table is highlighted ❷.

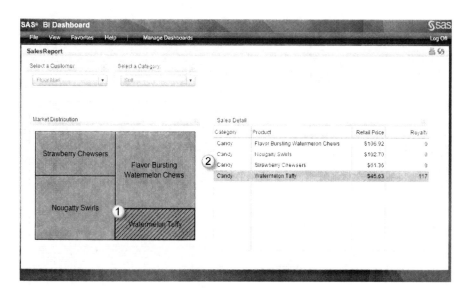

Figure 9.5-9 Second interaction example

9.6 SAS Administrator Tasks

Using SAS Management Console, the SAS administrator can set responsibilities and make system-wide changes that assist all users.

9.6.1 Roles and Responsibilities

Two user groups are defined in SAS Management Console: BI Dashboard Administrators and BI Dashboard Users.

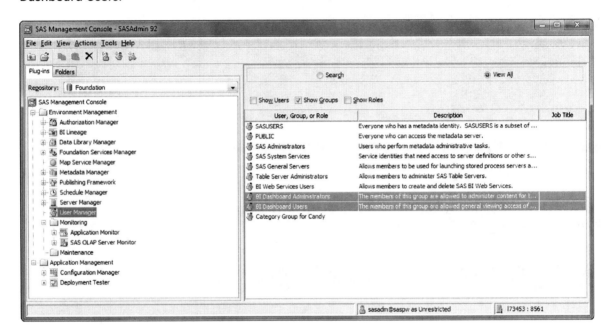

Figure 9.6-1 BI Dashboard roles

For users to have the ability to access the SAS BI Dashboard, they must be members of the BI Dashboard Users group. Developers tasked with creating and editing dashboard content must be members of the BI Dashboard Administrators group.

 Assign users to either the BI Dashboard Users group or the BI Dashboard Administrators group. Do not assign a user to the role BI Dashboard Administrator.

The BI Dashboard Administrator role defines what capabilities the BI Dashboard Administrators group has in other products, but in a default installation no capabilities are granted.

9.6.2 Data Caching

In version 4.3 of SAS BI Dashboard, data caching is not enabled by default. Each time a user opens a dashboard, each element on the dashboard executes a query to return the result. Enabling data caching can improve dashboard performance by generating data results in advance.

All data will be queried and cached using the SAS Trusted User identity. If your information maps for the dashboard include permissions based on user identity, then caching cannot be enabled.

You can choose which data sources can be enabled for caching by selecting one or more of the DSX files described in the table below.

DSX File Name	Related Data Source
dboard_sas.dsx	SQL query for data residing in the folder: <SAS Configuration Folder>\Lev1\AppData\SASBIDashboard4.3\sas-datasets
Infomap.dsx	Information maps
omr.dsx	Tables
stp.dsx	Stored process

Following are the steps to enable caching for a data table using the omr.dsx file.

1. In SAS Management Console, open **SAS Folders>System>Applications>SAS BI Dashboard>BI Dashboard 4.3>DataSourceDefinitions**.

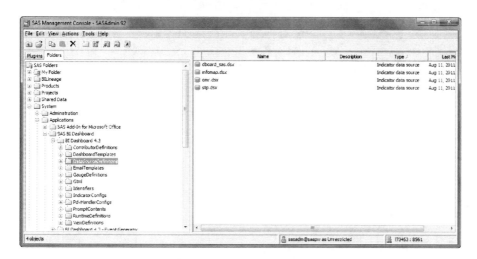

2. Right-click on the omr.dsk file and select **Write Content to External File**.

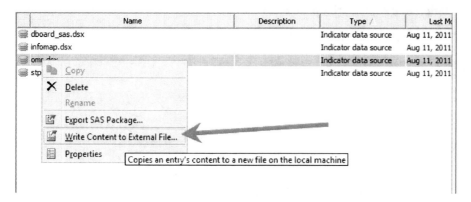

3. Save the file on your machine using the same name. In this example, the file name is omr.dsx.

4. Open the file in a text editor and remove the comments (<!-- and -->) around the
    DefaultTimingCacheDirective tag to enable caching and save your changes.

    Remove this text: <!--

                        <DefaultTimingCacheDirective cacheDisplayValueForRefresh="15.0"
                        cacheDisplayValueForStale="20.0"
                        cacheDisplayMultiplierForRefresh="MINUTES"
                        cacheDisplayMultiplierForStale="MINUTES"/>

    Remove this text: -->

5. Save the changes.

6. In SAS Management Console, right-click on the **DataSourceDefinitions** folder and select **Add
    Content From External File(s) or Directories**.

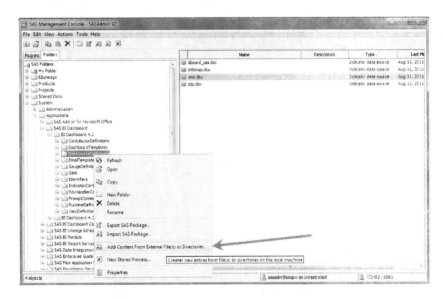

7. Select the DSX file you modified and replace the previous DSX file.

    Change the file type to **All Files** to see the omr.dsx file.

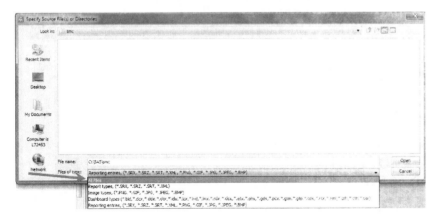

8. Confirm overwriting this file by selecting **Yes**.

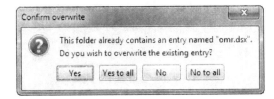

9. Restart the Web Application Server to see the changes in effect.

Go to support.sas.com for more information on enabling other data cache options for JDBC data sources.

9.6.3 Promotion Notes

Promoting dashboard content from development into production requires use of the Import/Export package from SAS Management Console to move metadata information. The different object types must be exported into separate packages (.spk files) and then imported into the target environment in the correct order. Following is the list of objects in the sequence required to complete a promotion.

1. Dependent objects

   Objects on which dashboard content depends must be promoted prior to beginning the promotion of the dashboard. These objects can include stored processes, data tables, OLAP cubes, and information maps.

2. Dashboard objects

   All of the developed dashboard objects will then need to be packaged and promoted. These can include the range, indicator data, indicator, and dashboard.

   The dashboard objects must be imported into the target system before the dashboard configuration files, otherwise some links between elements will not exist.

3. Dashboard configuration files

   Indicator configurations are stored within SAS Folders/System/Applications/SAS BI Dashboard/BI Dashboard 4.3/IndicatorConfigs.

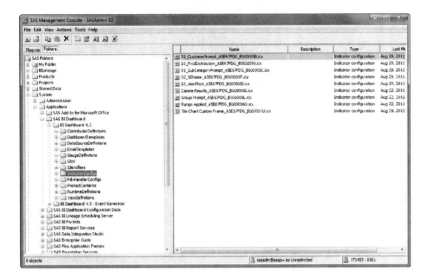

Figure 9.6-2 Indicator configurations within SAS Management Console

Prompt configurations are stored within SAS Folders/System/Applications/SAS BI Dashboard/BI Dashboard 4.3/PromptContents.

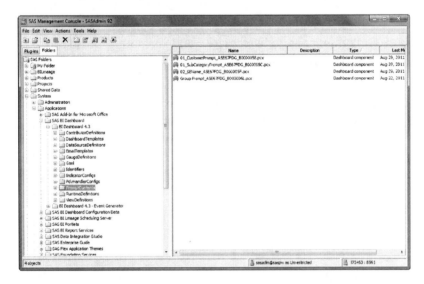

Figure 9.6-3 Prompt definitions found within SAS Management Console

 If users are allowed to create alerts in development and would like them promoted, these are stored in their user folder: SAS Folders/Users/<user id>/Application Data/BI Dashboard 4.3/IndicatorConfigs.

Chapter 10

SAS Information Delivery Portal

Chapter 10

SAS Information Delivery Portal

Bringing Everything Together

Every organization has its own focus and goals. As a result, one-size-fits-all reporting and data presentation tools do little to serve those organizations. The organization must decide what data is important, how data is reported, and how it is shared. Some organizations might have so many items to analyze that one single reporting technique will not suffice, while others might have a sole purpose but find it useful to understand all facets of the data via different techniques.

SAS Information Delivery Portal provides each organization with the flexibility to group, organize, and display the data and reporting in a single location. Using the portal, you can easily combine SAS reports, dashboards, stored process results, information maps, and even other Web-based applications in one location. Additionally, the security can also be addressed to ensure that the data is secure.

Many organizations define content administrators for each department to establish portal pages with specific reports and content access. Individuals might create personalized pages or portlets. These pages and portlets are then shared within the individuals' group or the entire organization.

In this chapter, you will learn how to set up a portal, create portlets, create custom layouts, and add customer content to the portlets.

10.1 Getting Started

As you begin learning about this tool, the following sections contain a review of the tool and what you need to get started.

10.1.1 Quick Tour

The following figure contains an example portal and explains the key display elements.

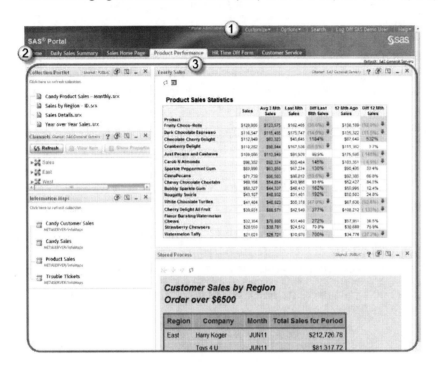

Figure 10.1-1 SAS Information Delivery Portal overview

1	Use the toolbar to add and customize pages, manage portlets, and control the page appearance.
2	Use the navigation bar to move between the portal pages. Each tab represents a portal page. Some content is preset by the content administrator. You can also build your own pages and add content based on your needs.
3	Use the portal pages to review the portlet contents. Portlets allow you to display multiple items on a page. In this example, there are three different portlets that show various data views.

10.1.2 Prerequisites

SAS Information Delivery Portal is Web-based and has a URL address similar to the following link. This link is case sensitive.

```
http://server name:port number/SASPortal
```

Your SAS administrator can give you the specific server address. If you are creating or viewing reports, you must have a user ID and password identified in the SAS system. Your profile determines which data you are allowed to view or use. If your organization needs to allow access to users who do not have access to the SAS system, refer to Section 10.6.2, "Accessing the Portal without Logging In."

 The portal supports only one user session per browser instance. Different portal users can have different sessions in separate browser instances, but not in separate tabs in the same browser instance.

10.2 Understanding Portlets

Each portal page contains one or more portlets. Each portlet allows you to add specific content to your page. Some portlets show a report from SAS Web Report Studio, while others might just be a list of information maps or external links. The following topics describe the more common portlets.

10.2.1 Using Report Portlets

Report portlets provide a way to display SAS Web Report Studio reports directly on the portal page. While a Collection portlet provides a link, this portlet allows you to display the report.

In Figure 10.2-1, the report is a simple horizontal bar chart. You will build similar reports in Chapter 7, "SAS Web Report Studio."

The two icons in the upper left of the window allow the user to control the data. The first icon refreshes the content. The second icon allows the user to view the report directly in SAS Web Report Studio.

Authorization access can limit what the user is able to view or change.

Figure 10.2-1 Report portlet

10.2.2 Using a Stored Process Portlet

One way to add a stored process to the portal is to use the Stored Process portlet. This portlet allows you to quickly display the output of a single stored process within the portal page. As shown in the following figure, the arrows allow you to navigate a large stored process and also refresh the data. This is useful when the report contains information that is updated on a regular basis.

Figure 10.2-2 Stored process portlet

10.2.3 Using a Collection Portlet

Collection portlets allow you to aggregate information. You might want to think of this portlet as a binder or a folder for various reports, links, documents, channels, and so on.

In Figure 10.2-3, a Collection portlet called Sales Reports is shown. This portlet contains SAS Web Report Studio reports, stored processes, an information map, a link to an intranet blog, and a publication channel. If you are interested in analyzing the quarterly candy sales, you do not have to search your e-mail, network drives, or the Internet. All data related to the topic has already been collected in one location.

After this content is created, it can be made available to users for their Home portal page or displayed in a portal page for the sales organization.

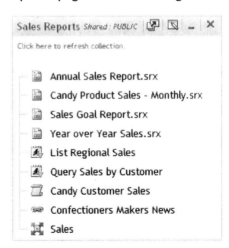

Figure 10.2-3 Collection portlet

10.2.4 Using the URL Portlet

Use the URL portlet when you want to display content such as external Web pages or even internal Web-based applications directly on the portal page.

You can link directly to stored processes from this portlet; however, authentication is not passed automatically from SAS Information Delivery Portal to SAS Stored Processes. Check Section 10.5.1.1, "Steps Required To Display Other SAS Content," for more information.

Figure 10.2-4 URL portlet

10.2.5 Using a Navigator Portlet

Using a Navigator portlet, you can see folder and subfolder contents that are in the metadata server. This portlet allows easy access to reports, information maps, stored processes, and packages.

As shown in Figure 10.2-5, the folder can be expanded to show the subfolders. To view an item, click the name to display its contents.

This portlet type is similar to the Collection portlet; however, this portlet would display all content within the specific area of the metadata folder structure. For example, the administrator can determine the starting folder and what content can be viewed. This could be set up to only allow the end user to see content in **My Folder**.

Using the Navigator portlet to display links to content that frequently change is better than adding the step of maintaining the Collection portlet.

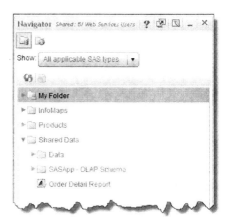

Figure 10.2-5 Navigator portlet

10.2.6 Using a Dashboard Portlet

Use the Dashboard portlet to share dashboards. Refer to Chapter 9, "SAS BI Dashboard," for more information about creating dashboards.

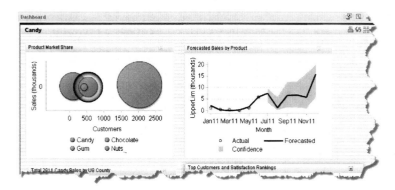

Figure 10.2-6 Dashboard portlet

10.2.7 Controlling Portlets

Each portlet has common elements and controls. The portlet contains the title, the group that can view the portlet, and two icons to control the portlet. The first icon (has an arrow expanding) allows you to edit the contents of the portal. If it is a Report portlet, you can select a report to add to it. If it is a Collection portlet, you can choose which links are included. (Refer to Section 10.3.3, "Adding a Report Portlet," to see how content is added.) The second icon (that has the pen and notepad) allows you to control the properties.

In the following figure, there are three different portlet banners so you can see how the sharing is handled. Two of the portlets are shared. The Collection portlet is shared with Sales Team –East, and the Sales Reports portlet is shared with the Public (everyone). The Stored Process 1 portlet does not have a shared assignment, which means the portlet is available only locally to the portal user and is not shared at all.

Figure 10.2-7 Sharing portlets

When the portlet is created, you can set the **Location (group)** option to share the portlet with other users. This can be changed later, so during the creation process, you might want to keep the portlet local until it is finished.

To update the sharing properties, do the following:

1.   Click the **Edit Properties** icon in the banner of the portlet you want to update.

2.   Select the group you want from the **Location (group)** drop-down menu. In the following figure, this portlet is set to **PUBLIC**, so it is now shared with everyone.

 You can select only one group to share a portlet or page with. If the portal groups are combinations of other groups, you should define portal-specific groups in SAS Management Console that are made up of these smaller groups.

10.3 Creating Your First Portal Page

Using SAS Information Delivery Portal, each portal page can address a specific organization need. In some cases, a new product release generates a large amount of attention. The entire organization might become focused on how the release is performing in the field and will be interested in all related data. One portal page that assembles all reporting and data would be beneficial and of high interest.

Other pages might be routine in nature, as in the case of a sales department manager who wants a page that allows her to track the sales performance. This is a simple page that contains two portlets. The first portlet is a predefined Collection portlet that one of your co-workers developed for the entire department. The second portlet is your manager's favorite report from SAS Web Report Studio.

10.3.1 Adding a New Portal Page

To start the creation process, you need to add and customize a portal page for the portlets you plan to add. Use the following steps to create a new portal page:

1. Open the SAS Portal window from your internet browser.

2. From the toolbar, select **Customize > Add page**.

3. In the **Create** tab, complete the fields as follows to add the page.

    a. In the **Name** field, add the page name. This page is what the user sees in the **Navigation** tab.

    b. In the **Description** field, add a brief description of the content. This field is not required, but it does help others understand whether they want to use the content.

    c. In the **Keywords** field, add words that assist users in finding the page.

    d. Use the **Page rank** field to order the tabs in the navigation window. By default, all new pages are assigned to rank 100, but if a specific order is required, the pages can be modified to follow a certain navigation order.

    e. The **Location (group)** drop-down list allows you to set the page permissions. This page is set to **PUBLIC** so that everyone can see it. Some pages might be limited to your group, in which case you might want to use another group, such as Sales Department.

    f. Select the **Share type** from the drop-down list.

Available	Users search for this page. It can be added or removed from view.
Default	Page is automatically added and user can remove it.
Persistent	Page is automatically added and cannot be removed from view.

4. Click the **Add** button and ensure that you see the system message. Then click the **Done** button.

   Note: If you do not first click the **Add** button, the change is not implemented and the page is not added. You will have to repeat the step.

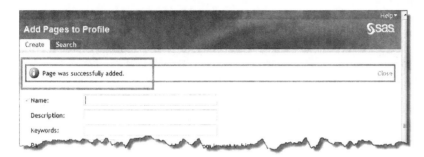

5. Confirm that the new portal page exists by reviewing the navigation bar. Because it was just created, the portal page is the last one in the row, as shown in the following figure.

10.3.2 Adding Previously Created Portlets

There might be common portlets in your organization that others find useful. Instead of trying to re-create content, you can use an existing portlet. In this example, another team member created a Collection portlet that contains links to sales reports and other related content that you think your manager can use when she needs to review the data.

To add a predefined portlet, do the following:

1. Go to the page you created in Section 10.3.1, "Adding a New Portal Page."

2. Select **Customize > Edit Page > Edit Page Content**.

3. The Edit Page Content window allows you to control the general page appearance. You are adding two portlets to this page. The Collection portlet is not large and does not need as much space as the Report portlet.

    Make the following changes to the page:

    a. **Layout** is **By column**. This lets the portlets use as much room in each column as needed.

    b. Change the **Number of columns** to **2**.

    c. Adjust the column width so Column 1 is 25% and Column 2 is 75%, as shown in the following figure.

    As you can see, there are several other combinations you could use to further customize your page. Refer to Section 10.4.1, "Changing the Layout," for more information about the layout.

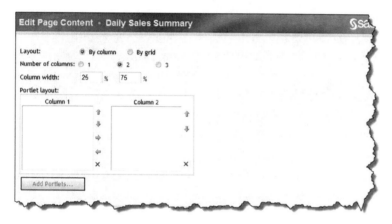

4. Click the **Add Portlets** button.

5. This content is predefined, so you need to search for it. Click the **Search** pane.

6. In the **Keywords** field, type a phrase to help you locate the portlet and click **Search**. The results are displayed in the right panel. Click the check box next to the portlet you want and click the **Add** button.

7. Confirm that the portlet was added. A message is displayed above the search results to confirm it was added. Otherwise, the portlet is not added and you will have to repeat the step.

8. Click the **Done** button to return to the Edit Page Content window. In the following figure, the portlet is displayed in the **Column 1** box.

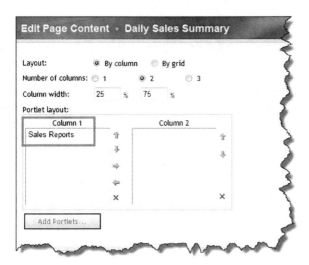

9. Click **OK** to return to the portal page. Because there are no other portlets on the page, this portlet uses all of the space, as shown in the following figure.

10.3.3 Adding a Report Portlet

The Report portlet displays SAS Web Report Studio reports. For this example, suppose that you created a Daily Sales Summary report earlier. You need to add the portlet and then assign the report to the portlet.

To add the report portlet, do the following:

1. Select **Customize > Edit Page > Edit Page Content**.

2. Select **Add Portlets** from the Edit Page Content window.

3. In the Add Portlets to Page window, complete the following fields to add the new portlet.

    a. In the **Portlet type** drop-down list, select **SAS Report Portlet**.

    b. In the **Name** field, type the portlet name. This is the name displayed to the user.

    c. In the **Location (group)** drop-down list, select **PUBLIC**.

    d. Select the **Add** button and confirm that the portlet was added.

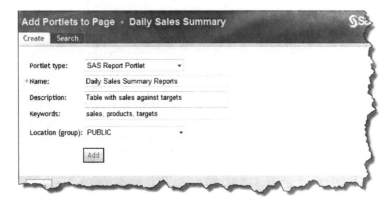

4. Select the **Done** button to return to the Edit Page Content window. The portlet defaults to the **Column 1** box, as shown in the following figure. Use the arrows to control the order and placement of the portlets.

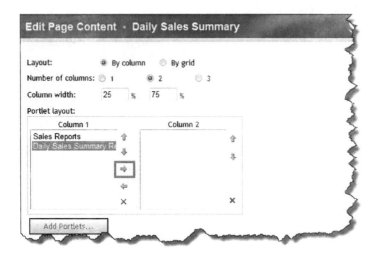

5. Move the Daily Sales Summary Report portlet to **Column 2**. Select **OK** to return to the page.

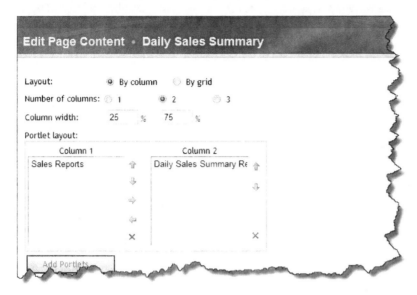

6. When the Report portlet is displayed initially, it does not have any content assigned to it. However, you can see that the SAS Portal window has two columns; Sales Reports is 25% of the display, while the Daily Sales Summary Reports has 75% of the display.

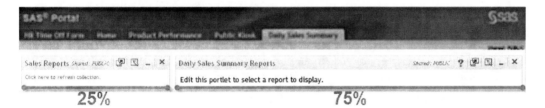

7. Click the **Edit Content** icon in the toolbar to select a report.

8.   In the Edit Content window, navigate to the report you want and select it. For this example, you want the report is called Daily Sales Summary.srx, which is stored in the Products library. Click **OK** to add the report.

The report is available in the SAS Portal page.

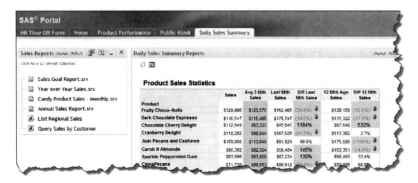

10.4 Enhancing Your Portal

You can change the portal layout and the page layouts to further personalize the portal look and feel.

10.4.1 Changing the Layout

You can control the portal page layout using the Edit Page Content settings. Using the Column and Grid layout controls, you can dramatically change the page appearance.

10.4.1.1 Using Columns Layout

This example uses one column that uses a Report portlet, as shown in the following figure. If you drew the layout on graph paper, the layout would look like a large block, represented by the blue square labeled 1 in the following figure.

The SAS Web Report Studio report, shown in Figure 10.4-1, was created to look like there are two columns; however, as you can see in the Edit Page Content window, the **Number of columns** field is set to **1** and **Report Portlet Used** appears in the **Column 1** box.

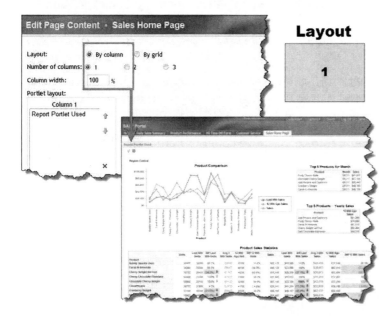

Figure 10.4-1 One-column layout

The next example uses three columns and multiple portlets. If you drew this layout on graph paper, it would look similar to the blue rectangles shown in the following figure. There would be three long columns, with the content *stacked* in each column.

In the following figure, the portlets are very different so you can more clearly see the layout. Each column is allotted 33% of the page. However, more likely content would be a stored process that showed a bar chart for each region.

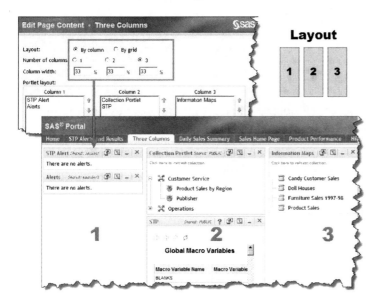

Figure 10.4-2 Three-column layout

10.4.1.2 Using Grid Layouts

The Grid layout allows you to have rows that each have columns of different sizes. If you drew this layout on graph paper, the layout shows Row 1 with Column 1 as a single column and Columns 2 and 3 combined; in Row 2, Columns 1 and 2 are combined and Column 3 is single column; and in Row 3, all columns are combined. (See the blue layout arrangement in the figure.)

 Portlet sizes are influenced by their content. If a Report portlet has large content, then it dictates how the other space in the portal is used. You might have to re-arrange the portlets until you get the layout that works for your content.

When specifying this layout, assign the same portlet to consecutive columns in the row. For instance, in Row 1, the Bookmarks portlet was assigned to Columns 2 and 3.

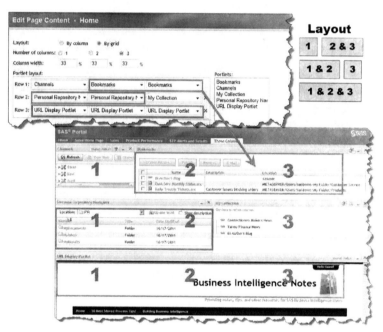

Figure 10.4-3 Three-column layout in Grid

10.4.2 Adjusting the Portlet Size

Some of the Collection portlets allow you to adjust their height. Click the **Edit Properties** icon on the portlet toolbar. Type the pixel size you want in the **Portlet height** field.

 Click the **Reset to default** check box to restore the default size.

Figure 10.4-4 Adjusting the portlet height

10.4.3 Changing the Page Tab Order

You can change the order of the page tabs and even the navigator bar location.

1. Select **Customize > Arrange Page Tabs**.

2. Change **Navigation bar location** to **Side**.

    You can also change the tab order by moving the arrows at the bottom of the list box. If the pages are ranked, click the **Use page rank** check box to place the pages in order by rank.

 You can add new pages using the **Add** button.

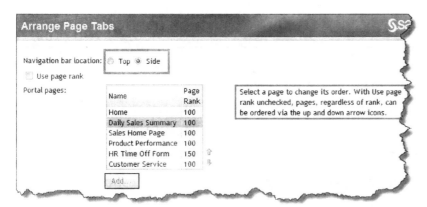

 When multiple pages are available for users and the page rank options for each page is used, click the **Use page rank** check box to display the page tabs in this order.

The following figure shows how the portal has changed after the modifications. The navigation bar is on the left side of the screen.

Note: You can move the navigation bar from horizontal to vertical placement by selecting **Preferences** from the **Options** drop-down menu.

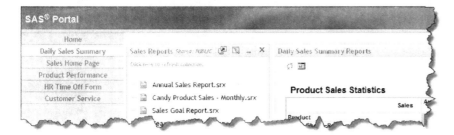

10.5 Tips and Tricks

Use the following tips and tricks to make your portal pages stand out.

10.5.1 Adding New Content

You can add new content to use in personal portal pages or to share with others in your organization. There are five different contents types that can be created:

Applications	Add Web-based applications such as internal, homegrown applications or other applications, such as Microsoft Outlook Express. Point to other SAS content such as stored processes, and address pass-through authentication between different SAS Web tools. See Section 10.5.1.1, "Steps Required To Display Other SAS Content," for more information.
Link	Add internal or external links to Web pages such as Google, CBS News, or blogs.
Syndication Channel	Add links to RSS feeds.
Portlet	Add a portlet.
Page	Add a page.

Use the following steps to set up a link to an external blog that can be shared with others or added as a bookmark.

1. Select **Options > Create New Content**.

2. Select the **Link** radio button.

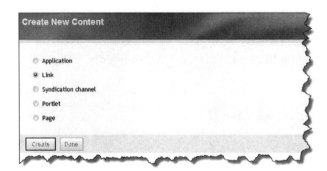

3. Complete the display as shown in the following figure to add a new external link.

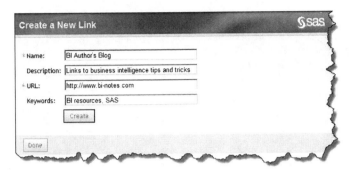

4. Select the **Create** button.

5. Select the **Done** button to exit the display. You can add the link to a collection portlet, to a URL portlet, or as a bookmark.

10.5.1.1 Steps Required To Display Other SAS Content

Authentication is not passed automatically between SAS components. If linking directly to stored processes, for example, the resulting window will show a second authentication screen prompting the user again for their user name and password. To remove this second authentication screen, you need to perform the steps described in Section 10.5.1, "Adding New Content."

To display the stored process screen on the portal page, the following steps are required.

1. Determine the full URL path from the SAS Stored Processes Web application.

2. Create an application, following the steps described in Section 10.5.1, "Adding New Content."

3. Search for the application from the portal.

4. Right-click on the application you just created and choose **Copy Hyperlink**.

5. Create a URL Display portlet for display on your portal page.

6. Edit the content of this URL Display portlet and paste the hyperlink you copied in step 4.

7. Save and display the portlet within the portal page. The output should include the stored process, rather than a second authentication window.

10.5.2 Adding Bookmarks

Your default portal page might contain a portlet for bookmarks. If not, you can add the portlet by searching for it. The Bookmark portlet allows each individual user to collect often used items in one location. Using the Bookmark portlet, you can create a portlet from a selected link, e-mail the bookmark, or even remove it from view.

Use the following information to search for bookmarks to add to your portlet:

1.   Select the **Search** menu item.

2.   In the **Keywords** field, type an asterisk (*) to see all the content available to you. You might want to limit your search to specific content types. For instance, in Section 10.5.1, you created a link to the SAS BI authors' blog. This link is the bookmark you want to locate.

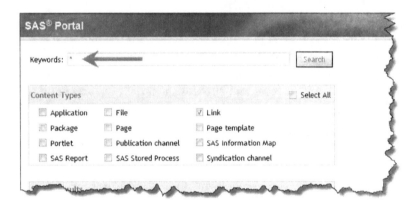

3.   The results are displayed in a new window. When you find the link you want, click the arrow icon and select **Bookmark** from the pop-up menu.

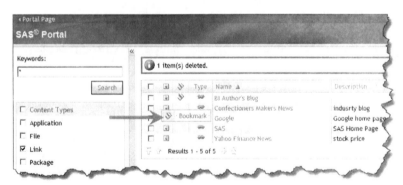

4. Confirm that the bookmark was added. Click **Portal Page** to return to the page.

The link is available in the portlet.

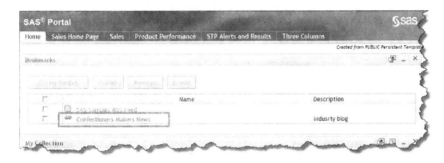

10.6 SAS Administrator Tasks

Using SAS Management Console, the SAS administrator can set responsibilities and make system-wide changes that assist all users.

10.6.1 Establishing Group Administrators

Group content administrators are responsible for managing portal pages shared with their group. These individuals are assigned to reduce the bottleneck that results when one SAS administrator completes all requests.

The following example creates the East Region Group and assigns a group administrator so that the team lead can manage the portal pages specific to this region.

1. Create the East Region Portal Group in SAS Management Console User Manager. Contact your SAS administrator if you need help (or permissions) to complete this step.

2. Add the team lead and all team members to this group.

3. Restart your Web server, which is usually JBOSS. SAS Information Portal does not create permission trees for new groups until it is restarted.

4. Within SAS Management Console, navigate to the permission tree and add the individual team lead to the **Authorization** tab.

   The permission tree is located in the Authorization Manager Plug-in under **Resource Management > By Type > Tree**.

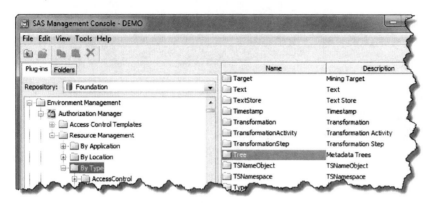

5. Double-click the East Region Portal Group Permissions Tree. Click the **Authorization** tab and select **Add**. Select the individual who will be responsible for administering the group.

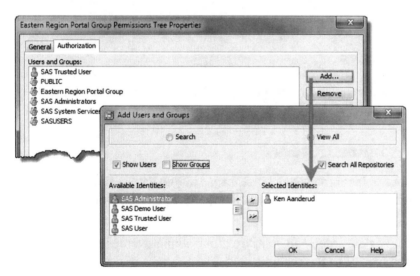

6. Grant the team lead WriteMetadata permission by selecting the check box.

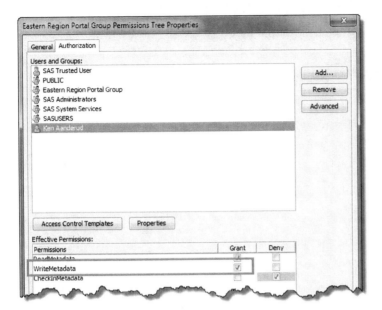

10.6.2 Accessing the Portal without Logging In

All users must first enter a user name and password to view portal pages and other content. In prior releases of SAS Information Delivery Portal, the Public Kiosk owned by internal SAS user SASGUEST was available to present users with a single screen that could include any content the organization wanted.

To allow users to view the content without logging in, you need to establish a new user account and enable it to act as the unchallenged user. Then portal users can directly view content through the following URL without logging into SAS.

```
<machine name>:<port number>/SASPortal/public
```

Remember that any users opening the application in this manner are accessing the system with this single new unchallenged account, which is called the Unchallenged Access User. Because of this, SAS administrators must be cognizant of not releasing secure content to the unchallenged user by mistake After they have implemented the unchallenged access, SAS administrators should complete an authentication audit of the SAS folder structure, the SAS Content Server, and SAS Information Delivery Portal content that is available but that links to other resources (such as dashboards or stored processes). You must ensure that only appropriate content is accessible by this new user.

1. Create an external account for the unchallenged access user. This account will be used for creating and including content on the public portal page.

2. Reconfigure SAS Information Delivery Portal through SAS Management Console to allow the unchallenged user.

    Open the application properties from the configuration manager.

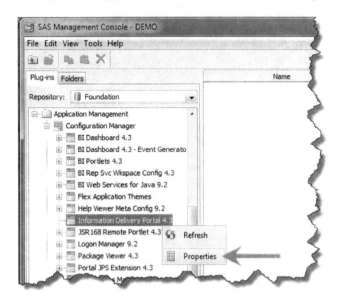

3. In the **Advanced** tab, edit all four values for **Unchallenged.Access** choices according to your organization's requirements. In the following example, you want to hide the **LogIn**, **Logoff**, and **Search** buttons for the unchallenged users.

    a. In the **Unchallenged.Access.Enabled** field, type `true`.

    b. In the **Unchallenged.Access.Logoff.Behavior** field, type `hide`.

        Additional options for Logoff.Behavior include:

        *   **logoff** shows the term **Logoff**. Clicking **Logoff** forwards users to the Logoff message window.

        *   **hide** does not display **Login** or **Logoff**.

    c. In the **Unchallenged.Access.Show.Search.Menu** field, type `false`.

        If you leave this value as `true`, which is the default, the **Search** link allows unchallenged users to search for other content to view.

Typically, you will set the value to `false` to ensure that content not intended for the unchallenged user remains unavailable.

d. In the **Unchallenged.Access.UserID** field, type the user account ID that you created in step 1.

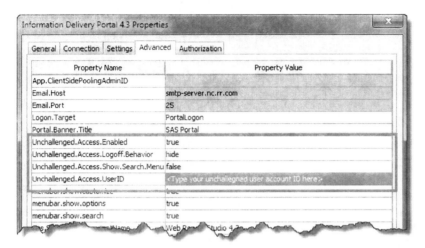

4. Redeploy the SAS Information Delivery Portal web application by running Deployment Manager. Deployment Manager is located in the program files folder. In typical Windows installations, this is C:\Program Files\SAS\SASDeploymentManager\9.2\config.exe.

5. Select the **Rebuild Web Applications** button and follow the wizard through creating a new portal Web deployment.

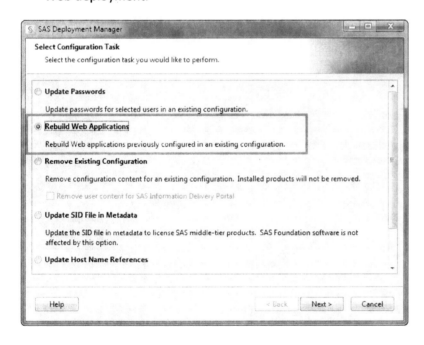

6. After the application is rebuilt, log into the SAS Information Delivery Portal page with the unchallenged user account created in step 1. Create content for the SASPortal/public page as you would for yourself.

Since you are creating this as the end user that will view the content, you do not need to change the **Location** option for this content.

Index

CPSIA information can be obtained at www.ICGtesting.com
Printed in the USA
LVOW10s0811201114

414607LV00001B/4/P